Praise for *The Biotech C*

"*The Biotech Century* is Rifkin's latest and most
the pace at which genetic engineering and the c
view, doom humanity unless more public aware[illegible] of this ominous trend increases. . . . He has filled this book with an astonishing range of research published in scientific journals and in the popular media. He is a brilliant thinker and a powerful writer . . . a shrewd debater in public and in print."

—*San Francisco Chronicle*

"Jeremy Rifkin has written a most remarkable book on the coming Biotech Century, full of information that, as far as I know, has for the first time been collected in such completeness. It deserves to be read by everybody, be he or she optimist or pessimist."

—Erwin Chargaff, professor emeritus of biotechnology, Columbia University, and discoverer of "Chargaff's Rules"

"Though he does not dispute the promised benefits of biotechnology, Rifkin warns that we must closely consider its possible (and often little-publicized) negative consequences. . . . Wide-ranging and deeply intelligent."

—*Publishers Weekly*

"It is probably best to make clear straight away that this book is not, or not just, an attack on genetic engineering. Jeremy Rifkin is struck by the change that the whole process and its prospects represent in our attitude to nature—and consequently in the way we conceive of ourselves . . . this is a shrewed, helpful, and far-sighted book."

—*New Scientist*

"Jeremy Rifkin, as usual, is ahead of his time. For twenty years, he's been warning about the hazards of genetic engineering. The rest of the world acts like it was an issue born yesterday."

—*Business Ethics*

Also by Jeremy Rifkin

*Common Sense II*
*Own Your Own Job*
*Who Should Play God?* (with Ted Howard)
*The Emerging Order*
*The North Will Rise Again* (with Randy Barber)
*Entropy* (with Ted Howard)
*Algeny*
*Declaration of a Heretic*
*Time Wars*
*Biosphere Politics*
*Beyond Beef*
*Voting Green* (with Carol Grunewald)
*The End of Work*

Harnessing
the Gene
and Remaking
the World

# The Biotech Century

Jeremy Rifkin

Jeremy P. Tarcher / Putnam
a member of
Penguin Putnam Inc.
New York

Most Tarcher/Putnam books are available at special quantity discounts for bulk purchases for sales promotions, premiums, fund-raising, and educational needs. Special books or book excerpts also can be created to fit specific needs. For details, write Putnam Special Markets, 375 Hudson Street, New York, NY 10014.

Jeremy P. Tarcher/Putnam
a member of
Penguin Putnam Inc.
375 Hudson Street
New York, NY 10014
www.penguinputnam.com

First trade paperback edition 1999

Published simultaneously in Canada

Library of Congress Cataloging-in-Publication Data
Rifkin, Jeremy.
The biotech century : harnessing the gene and remaking the world /by Jeremy Rifkin.
p. cm.
"A Jeremy P. Tarcher/Putnam book."
ISBN 0-87477-953-7 (acid-free paper)
1. Biotechnology—Social aspects. 2. Biotechnology—Moral and ethical aspects. 3. Genetic engineering—Social aspects. 4. Genetic engineering—Moral and ethical aspects. I. Title.
TP248.2.R54 1998
303.48'3—dc21 97-44358 CIP

*Book design by Ralph L. Fowler*
*Jacket design by Lee Fukui*

Printed in the United States of America
1 2 3 4 5 6 7 8 9 10
This book is printed on acid-free paper. ♾

# Acknowledgments

I would like to thank Jon Akland, who assisted me as researcher for *The Biotech Century.* His painstaking research and relentless attention to detail were invaluable in the preparation of the manuscript. I'd also like to thank Stuart Newman, Martin Teitel, and Ted Howard for reading early drafts of the book and providing so many useful suggestions and comments for improving the work; Anna Awimbo for her many constructive editing changes, which helped guide the book through its various stages; and Clara Mack and Joyce Wooten, who both were immensely helpful in the preparation of the manuscript.

I'd like to give special thanks to my longtime friend Jeremy Tarcher, and my editor, Mitch Horowitz, whose keen editing assistance helped take the many ideas in this book and turn them into a story line. I would also like to extend my personal thanks to my publisher, Joel Fotinos, for his strong stewardship and support of this book from its inception.

Finally, I'd like to thank my wife and fellow writer, Carol Grunewald, whose many editorial suggestions were incorporated into virtually every page of the book, helping turn a work in progress into a finished book. I'd also like to thank my wife for suggesting that I write this book on the eve of the Biotech Century. Her persistence is largely responsible for this effort.

For my nieces and nephews,

David, Susan, Gregory,

Michael, Jonathan, Dayna,

Rachel, Traci, and Mark,

and all the members

of the next generation

who will face the many

opportunities and challenges

of the Biotech Century

# Contents

# Introduction

It was more than twenty years ago that I co-authored, with Ted Howard, a book entitled *Who Should Play God?* In that book we wrote of the promises and perils of a fledgling new technology that few people had heard of called genetic engineering. While we discussed the many benefits that would result from the new science, we also warned of the dangers that might accompany the new technology revolution. Among other things, we predicted that transgenic species, animal chimeras and clones, test tube babies, the rental of surrogate wombs, the fabrication of human organs, and human gene surgery would all be realized before the end of the current century. We also said that screening for genetic diseases would become widespread, raising serious questions about genetic discrimination by employers, insurance companies, and schools. We expressed our concern over the increasing commercialization of the Earth's gene pool at the hands of pharmaceutical, chemical, and biotech firms, and raised questions about the potentially devastating long-term impacts of releasing genetically engineered organisms into the environment.

At the time, the nation's molecular biologists, policy leaders, media pundits, and editorial and science writers dismissed our predictions as "alarmist" and far-fetched, and argued that the science we wrote about was at least a hundred years away, perhaps several hundred years away. The conventional wisdom among most scientists working in the new field was that there was little need to examine the environmental, economic, social, and ethical implications of what they claimed was a "hypothetical" future.

In the twenty years that have elapsed since the publication of that book, every scientific and technological breakthrough we predicted has occurred. Yet, sadly, leaders in the scientific community, the media, the government, and the business community, with a few notable exceptions, remain as reluctant today as they were a generation ago to engage in a broad public debate over what is likely to be the most radical experiment humankind has ever carried out on the natural world. I note this not out of rancor but, rather, to set the record straight on how this science has unfolded in the public arena. It is true that with each announcement of a major breakthrough in the new science of genetic engineering, the press has devoted a small amount of copy to reporting the "other side of the story." It has usually been in the form of a critical aside or afterthought, often buried deep inside the story, in an attempt to lend a faint air of objectivity to the coverage.

On occasion, the press has delved more deeply into some of the worrisome issues raised by the new technologies, especially with regard to genetic screening and discrimination. By and large, however, the public has been subjected to an unending flow of glowing, and largely uncritical, reports of the many new breakthroughs on the genetic frontier, with little effort to examine the more complex risks, pitfalls, and dangers that accompany the biotech revolution—issues that cry out for public airing as we turn the corner into the century of biology. If anyone cares to survey the two decades of reporting on the biotech revolution, in the trade, the business press, and general media, they will be able to see for themselves how much print and air time has been devoted to the claims of the geneticists and the biotech industry and how little time has been given over to legitimate issues of concern raised by the growing number of critics.

Rarely have serious questions relating to the risks of the new science and its commercial application been raised by those financing or conducting the research. The one exception that comes to mind was a short-lived self-imposed moratorium on gene-splicing research by molecular biologists in the mid-1970s, pending the establishment of federal guidelines to ensure proper safety protocols in the nation's biotech labs. Even here, however, scientists were motivated more by issues of personal and institutional liability than by concern for the potential environmental consequences of gene-spliced organisms escaping into the environment.

On July 26, 1974, eleven of the leading scientists in the new field of molecular biology published an open letter asking their colleagues to initiate a self-imposed moratorium on conducting high-risk recombinant DNA experiments, to allow time for a discussion of the potential safety issues involved in the new research. At a follow-up conference held at Asilomar, California, in February of 1975, one hundred and forty biologists from seventeen countries met to consider the environmental and health risks of conducting recombinant DNA experiments. The science press reported that many, if not most, of the attendees at the meeting were anxious to move ahead with their work and opposed to any regulation of their research. An article in *Science News* said that the scene at Asilomar was one of "unyielding, self-indulgent and conflicting attitudes."[1] By the conference's end, it began to appear that the moratorium would be lifted. The mood changed abruptly, however, on the third day when several attorneys gave presentations on the legal responsibility of researchers who create a "bio-hazard." The final speaker, Professor Harold Green of the George Washington University Law School, caught the attention of the participants with a talk entitled, "Conventional aspects of the law, and how they may sneak up on you in the form, say, of a multi-million–dollar lawsuit."[2]

Financial self-interest won out. The next morning, at the final session of the conference, the assembled scientists endorsed a two-point safety program. They established broad, general categories of experiments that represented differing degrees of risk, and designated the type of laboratory precautions they believed would provide physical containment of potential hazards. They also developed the novel concept of "biological containment" by agreeing to use an enfeebled *E. coli* bacterium as a host for their recombinant DNA chimeras. The use of this biologically weak host, scientists hoped, would prevent any chimera from living in the natural environment. The risk classification scheme was subsequently redefined by the National Institutes of Health and drafted into detailed guidelines that established four levels of risk and the safety requirements to meet them.

A few courageous molecular biologists, and a greater number of ecologists, have spoken up over the years, expressing concerns over some of the more troubling aspects of the emerging biotech revolution, often at great risk to their own careers. They are, to my mind, unsung heroes. Whenever criticism has been raised, doubts expressed, and misgivings shared, however, the response by those in the "scientific establishment" and in industry has been less than generous. Genuine debate has been marginalized and, on occasion, stifled by an establishment that views every question, query, and reservation as a direct and immediate assault on free inquiry and science itself. Critics, myself included, have been attacked over and over again as

Luddites, vitalists, fearmongers, and fundamentalists for broaching concerns over where the new science is heading, the implication being that any questioning of the "conventional wisdom" is heresy or, even worse, lunacy.

My hope, from the very beginnings of this new science, is that we might learn the lessons of the two earlier scientific revolutions in physics and chemistry in the nineteenth and twentieth centuries. Both brought great benefits to humankind along with equally significant problems. Perhaps, had previous generations been more willing to entertain a thoughtful, open, public debate over the potential risks as well as rewards of these two scientific revolutions, early on, before they ran their course, we, and more importantly, our children, might not be so saddled with the growing environmental, social, and economic bill that is also the shared legacy of the modern age.

Although the window is rapidly closing, we still have an opportunity to raise some of the tough issues, up front, regarding the new biotechnology revolution. Let me also say, at the outset, that the many issues surrounding this new technology revolution are complex and not easily reducible to an either/or response. Were it just a matter of condemning a monstrous and unwarranted intrusion into our lives, a short polemic would suffice to lay bare the "terrible nature" of the new science that is descending upon us. But this new science is far more complicated than that. Genetic engineering represents our fondest hopes and aspirations as well as our darkest fears and misgivings. That's why most discussions of the new technology are likely to be so heated. The technology touches the core of our self-definition. The new tools are the ultimate expression of human control—helping us shape and define the way we would like to be and the way we would like the rest of living nature to be. Biotechnologies are "dream tools," giving us the power to create a new vision of ourselves, our heirs, and our living world and the power to act on it.

I have devoted considerable time to chronicling the host of new biotech products and processes being readied for the marketplace, as well as providing detailed accounts of the many feats at the cutting edge of this new technology revolution. I've also given time over to the claims—some of which are no doubt exaggerated and self-serving—being made on behalf of the new science by its champions and advocates in the research and business communities. When a colleague read an early draft of this book he commented that the list of particulars cited in the manuscript sounded so inviting that he was left wondering whether I was making the case for the new biotechnologies. My answer was yes and no. Who wouldn't approve of some of the extraordinary advances coming out of the biotech laboratories? The new gene-splicing technologies address so many of our yearnings and desires. They promise a better way of life. Some, but not all, of the new products and services will even deliver on those promises. Only the most committed contrarian would say, "I see no value in the new science and technology." In fact, there is value, great value, in some of the products of genetic engineering and that's what makes the discussion of this ultimate human technology so interesting, difficult, and challenging.

On the other hand, the new genetic science raises more troubling issues than any other technology revolution in history. In reprogramming the genetic codes of life, do we risk a fatal interruption of millions of years of evolutionary development? Might not the artificial creation of life spell the end of the natural world? Do we face becoming aliens in a world populated by cloned, chimeric, and transgenic creatures? Will the creation, mass production, and wholesale release of thousands of genetically engineered life forms into the environment cause irreversible damage to the biosphere, making genetic pollution an even greater threat to the planet than nuclear and petrochemical pollution? What are the consequences for the global economy and society of reducing the world's gene pool to patented intellectual property controlled exclusively by a handful of multinational corporations? How will the patenting of life affect our deepest convictions about the sacred nature and intrinsic value of life? What is the emotional and intellectual impact of growing up in a world where all of life is treated as "invention" and "commercial property"? What will it mean to be a human being in a world where babies are genetically designed and customized in the womb and where people are identified, stereotyped, and discriminated against on the basis of their genotype? What are the risks we take in attempting to design more "perfect" human beings? I've used this book to explore these issues and some of the many other critical concerns that need to be raised before we lock ourselves into a genetically engineered future. The new genetic engineering technologies being readied for the commercial market need to be fully scrutinized in the coming years if we are to minimize the risks to future generations and the other creatures who travel with us on life's journey.

The Biotech Century comes to us in the form of a grand Faustian bargain. We see before us the lure of great strides and advances and a bright future full of hope. But, with every step we take into this "Brave New World," the nagging question "At what price?" will haunt us. This book, more than anything else, is about value and cost, gains and losses, on both a personal and worldwide scale. The risks attendant to the Biotech Century are at least as ominous as the rewards are seductive. Wrestling with the light and dark sides of biotechnology will test each of us in our own way.

Some readers will, I suspect, finish this book convinced that, on the whole, the benefits outweigh the risks, and favor proceeding with the genetic engineering of life. Others will likely oppose much or all of this new technology revolution. There will also be those in the middle, hopeful yet troubled, conflicted and unsure of how they feel about the marvels and mishaps that are likely to accompany our journey into the Biotech Century. Then, too, positions are not cast in stone. Developments in the new science and technology and in the global economy and society in the ensuing years could harden, temper, or weaken the resolve of previously held views. For now, the most important issue at hand is to make the new science and technology an issue of considerable public attention.

We need to open the doors up wide to a thoughtful public dialogue of the pros and cons of the new biotechnologies. We also have to understand, in the process of that discussion, that genetic engineering technology is not a fait accompli, but merely one option among many. Human beings make history and we do so by making choices among competing possibilities and priorities. The notion of free exercise of will is fundamental to any serious discussion of the new technologies. The question, in the final analysis, is not how do we learn to live with the new genetic technologies, but whether, and under what condition, we want them to be a part of our lives. If we are not willing and able to entertain the notion of a measured rejection as well as a qualified acceptance of the new technologies, then what we regard as debate will really be little more than a handwringing exercise and of little consequence. After all, if the technology is deemed inevitable, then any real debate is useless.

I have struggled with the question of how much I should inject my personal views into the book, and how much I should let the information speak for itself. Some readers might feel that they didn't hear enough from me on my own ideas; others may think that I spent too much time presenting my opinions. While I've certainly not made any attempt to hide my own biases and feelings, which are reflected in personal commentary throughout the book, I wanted to leave some breathing room for the reader, some open space amidst the swirl of claims and counter-claims for personal rumination on the issues at hand. I hope that balance is reflected in the pages that follow. Of this much I think we can all agree. The new genetic engineering technologies being introduced into society will intimately affect our lives, requiring each and every one of us to become engaged, at some level, in a discussion of the values at stake.

I have written this book, on the eve of what many in the scientific and business community are calling "the Biotech Century," because of two striking developments. First, the genetic revolution and the computer revolution are just now coming together to form a scientific, technological, and commercial phalanx, a powerful new reality that is going to have a profound impact on our personal and collective lives in the coming decades. Second, many of the scientific breakthroughs we predicted more than twenty years ago are now moving out of the laboratory and into widespread commercial use, bringing us face to face, for the first time, with the promises and perils of this "new era." We are past due for an informed, sober debate on the many issues raised by the biotech revolution.

I have chosen not to examine the many benefits of introducing alternative biotechnologies—those emphasizing sustainable ecological practices, including organic agriculture, solar power, and preventive medicine. I have, however, made brief reference to this other vision of the Biotech Century in chapter eight. A thoughtful and thorough examination of this other approach to the Biotech Century would require a separate analysis and a book of at least equal length to do justice to the topic.

In a few instances, I have revised and updated material from past editorials, articles, essays, and books I have authored that remain relevant to the current debate. Chapter seven, "Reinventing Nature," originally appeared as part of an essay I wrote more than fifteen years ago. I had written the essay in anticipation of the coming together of the information and life sciences into a single scientific, technological, and commercial revolution. At that time, however, bioinformatics had not yet been developed and computers and software technology were playing only a tangential role in the unfolding biotech revolution. Now, all that has changed. The marriage of computers and genetic science, in just the last ten years, is one of the seminal events of our age and is likely to change our world more radically than any other technological revolution in history. I have revised and made substantial changes in my earlier essay on "Reinventing Nature" and included it in the book as it is even more relevant to the public discussion today than when originally written.

We are entering a new century and a new millennium full of promise and expectations, tempered by growing concerns and doubts. I hope this book will contribute to a thoughtful discussion of the possibilities and options that lie ahead.

*Jeremy Rifkin*
*Washington, D.C.*
*January 10, 1998*

# One The Biotech Century

Never before in history has humanity been so unprepared for the new technological and economic opportunities, challenges, and risks that lie on the horizon. Our way of life is likely to be more fundamentally transformed in the next several decades than in the previous one thousand years. By the year 2025, we and our children may be living in a world utterly different from anything human beings have ever experienced in the past.

In little more than a generation, our definition of life and the meaning of existence is likely to be radically altered. Long-held assumptions about nature, including our own human nature, are likely to be rethought. Many age-old practices regarding sexuality, reproduction, birth, and parenthood could be partially abandoned. Ideas about equality and democracy are also likely to be redefined, as well as our vision of what is meant by terms such as "free will" and "progress." Our very sense of self and society will likely change, as it did when the early Renaissance spirit swept over medieval Europe more than seven hundred years ago.

There are many convergent forces coming together to create this powerful new social current. At the epicenter is a technology revolution unmatched in all of history in its power to remake ourselves, our institutions, and our world. Scientists are beginning to reorganize life at the genetic level. The new tools of biology are opening up opportunities for refashioning life on Earth while foreclosing options that have existed over the millennia of evolutionary history. Before our eyes lies an uncharted new landscape whose contours are being shaped in thousands of biotechnology laboratories in universities, government agencies, and corporations around the world. If the claims already being made for the new science are only partially realized, the consequences for society and future generations are likely to be enormous. Here are just a few examples of what could happen within the next twenty-five years.

A handful of global corporations, research institutions, and governments could hold patents on virtually all 100,000 genes that make up the blueprints of the human race, as well as the cells, organs, and tissues that comprise the human body. They may also own similar patents on tens of thousands of micro-organisms, plants, and animals, allowing them unprecedented power to dictate the terms by which we and future generations will live our lives.

Global agriculture could find itself in the midst of a great transition in world history, with an increasing volume of food and fiber being grown indoors in tissue culture in giant bacteria baths, at a fraction of the price of growing staples on the land. The shift to indoor agriculture could presage the eventual elimination of the agricultural era that stretched from the neolithic revolution some ten thousand years ago, to the green revolution of the latter half of the twentieth century. While indoor agriculture could mean cheaper prices and a more abundant supply of food, millions of farmers in both the developing and developed world could be uprooted from the land, sparking one of the great social upheavals in world history.

Tens of thousands of novel transgenic bacteria, viruses, plants and animals could be released into the Earth's ecosystems for commercial tasks ranging from "bio-remediation" to the production of alternative fuels. Some of those releases, however, could wreak havoc with the planet's biosphere, spreading destabilizing and even deadly genetic pollution across the world. Military uses of the new technology might have equally devastating effects on the Earth and its inhabitants. Genetically engineered biological warfare agents could pose as serious a threat to global security in the coming century as nuclear weapons do now.

Animal and human cloning could be commonplace, with "replication" partially replacing "reproduction" for the first time in history. Genetically customized and mass-produced animal clones could be used as chemical factories to secrete—in their blood and milk—large volumes of inexpensive chemicals and drugs for human use. We could also see the creation of a range of new chimeric animals on Earth, including human/animal hybrids. A chimp/hume, half chimpanzee and half human, for example, could become a reality. The human/animal hybrids could be widely used as experimental subjects in medical research and as organ "donors" for xenotransplantation. The artificial creation and propagation of cloned, chimeric, and transgenic animals could mean the end of the wild and the substitution of a bioindustrial world.

Some parents might choose to have their children conceived in test tubes and gestated in artificial wombs outside the human body to avoid the unpleasantries of pregnancy and to ensure a safe, transparent environment through which to monitor their unborn child's development. Genetic changes could be made in human fetuses in the womb to correct deadly diseases and disorders and to enhance mood, behavior, intelligence, and physical traits. Parents might be able to design some of the characteristics of their own children, fundamentally altering the very notion of parenthood. "Customized" babies could pave the way for the rise of a eugenic civilization in the twenty-first century.

Millions of people could obtain a detailed genetic readout of themselves, allowing them to gaze into their own biological futures. The genetic information would give people the power to predict and plan their lives in ways never before possible. That same "genetic information," however, could be used by schools, employers, insurance companies, and governments to determine educational tracks, employment prospects, insurance premiums, and security clearances, giving rise to a new and virulent form of discrimination based on one's genetic profile. Our notions of sociality and equity could be transformed. Meritocracy could give way to genetocracy, with individuals, ethnic groups, and races increasingly categorized and stereotyped by genotype, making way for the emergence of an informal biological caste system in countries around the world.

The Biotech Century could bring some or even most of these changes and many more into our daily lives, deeply affecting our individual and collective consciousness, the future of our civilization, and the biosphere itself. The benefits and perils of what some are calling "the ultimate technology frontier" are both exciting to behold and chilling to contemplate. Still, despite both the formidable potential and ominous nature of this extraordinary technology revolution, until now far more public attention has been focused on the other great technology revolution of the twenty-first century—computers and telecommunications. That's about to change. After more than forty years of running on parallel tracks, the information and life sciences are slowly beginning to fuse into a single technological and economic force. The computer is increasingly being used to decipher, manage, and organize the vast genetic information that is the raw resource of the emerging biotech economy. Scientists working in the new field of "bioinformatics" are beginning to download the genetic information of millions of years of evolution, creating a powerful new genre of "biological data banks." The rich genetic information in these biological data banks is being used by researchers to remake the natural world.

The marriage of computers and genes forever alters our reality at the deepest levels of human experience. To begin to comprehend the enormity of the shift taking place in human civilization, it's important to step back and gain a better understanding of the historic nature of the many changes that are occurring around us as we turn the corner into a new century. Those changes represent a turning point for civilization. We are in the throes of one of the great transformations in world history. Before us lies the passing of one great economic era and the birth pains of another. As the past is always prelude to the future, our journey into the Biotech Century needs to begin with an account of the world we're leaving behind.

## The End of the Industrial Era

The industrial saga is coming to an end. It was a singular moment in world history characterized by brawn and speed. We burrowed below the Earth where we found millions of years of stored sunlight in the form of coal, oil, and natural gas—a seemingly unlimited source of energy that could be used, with the aid of the steam engine, and later the electrodynamo, to speed the delivery of what we called material progress.

Our sense of place and space was fundamentally changed. Millions of human beings around the world were uprooted from their rural homes and ancestral lands and forced to make their way into sprawling new urban enclaves where they sought a new kind of work in dimly lit factories and bustling offices far removed from the changing seasons and the age-old customs and rituals of an agricultural existence.

Rails were laid across continents, followed in quick succession by telegraph and telephone wires and poles and miles of paved roads, changing our notions of time and distance. The Wright brothers took wings to flight and a few years later Henry Ford began supplying every American family with a standardized gasoline-powered automobile. Time zones and posted speeds heralded a new quickened pace of life, and in schools, businesses, and homes, the talk was of efficiency, the new mantra of a streamlined, hard-hitting, "future oriented" century.

Everywhere, cement foundations were laid and scaffolding went up, making way for a new vertical world of gleaming secular cathedrals made of iron, steel, aluminum, and glass. After centuries of working, living, and socializing side by side, we began a radical new experiment, living in relative isolation, one on top of another, always in search of that elusive prize which, for want of a better name, we loosely called self-fulfillment.

It was an age of great abundance for some and growing destitution for others—giant department stores, and later, vast shopping malls were filled with exotica and trivia, staples, craft, and fine works of art. Fashion replaced necessities for millions of people.

Scientists and engineers became our new authorities on almost everything that mattered, their views and ideas sought out, revered, elevated, and enshrined. It was the century of physics and chemistry. We peered into the micro world of atoms and electrons and rewrote the book of nature with the discoveries of quantum mechanics and relativity theory. Scientists split the atom, harnessed a new form of energy more powerful than anything human beings had ever experienced, and created the atomic bomb. Physicists and engineers took man to the moon and back while chemists busied themselves with the creation of new kinds of more versatile synthetic materials. Plastics, a curiosity at the beginning of the century, became ubiquitous by the mid decades, seemingly as essential to our way of life as the very air we breathe. Petrochemical fertilizers and synthetic pesticides reshaped the agricultural landscape, coming just in time—claim many advocates—to help feed a growing human population that was doubling every two generations.[1]

We built vast sewer systems underground, aerated and purified our water, improved our nutrition and increased our life expectancy by more than twenty years. Engineers invented the X-ray machine and later the MRI. Medical researchers gave us vaccinations, anesthetics, antibiotics, and other wonder drugs.

At long last, we are nearing the end of this unique period in world history—the industrial era spread across five centuries and six continents and fundamentally changed the way human beings lived, worked and viewed themselves and their world. It was, above all else, an age propelled by cheap and abundant extractive energy. Once regarded as nearly inexhaustible, the fuel reserves of the carboniferous era are steadily diminishing, making it more costly to extract and more expensive to consume. Watching the Persian Gulf War on television night after night as hundreds of oil derricks across the Kuwaiti desert poured fire into the sky, consuming millions of barrels of precious oil for weeks and then months at a time, was a powerful reminder of just how dependent the modern world has become on these precious fuels.

At the same time that our energy reserves are dwindling, the entropy bill for the Industrial Era is coming due, forcing us to look at the red ink on the modern ledger. Our affluence has been purchased at a steep price. The spent energy of hundreds of years of burning fossil fuels has begun to accumulate in the form of increasing greenhouse gases in the biosphere. Global warming is changing the very biochemistry of the planet, threatening temperature and climate changes of incalculable proportions in the coming century. Even a three-and-one-half degree Fahrenheit change in temperature brought on by global-warming gases—regarded by most scientists as a conservative forecast of what might be in store—would represent the most significant change in the Earth's climate in thousands of years. A climate change of this magnitude will likely lead to the melting of the polar ice caps, a worldwide rise in sea water level, the submerging of some island nations, the erosion of coastlines, and radical fluctuations in weather including more severe droughts, hurricanes, and tornadoes. Whole ecosystems are likely to fall victim to the radical climate shift. Agricultural regions are likely to shift far to the north, creating new opportunities for some and loss of livelihood for others.

The steady decline in fossil fuel reserves and the increasing global pollution brought on, at least in part, by the use of these same fuels is leading civilization to search for new, alternative approaches to harnessing the energy of nature in the coming century. The hard reality is that we are nearing the end of the age of fossil fuels and, with it, the end of the industrial age that has been molded from it. While the Industrial Age is not going to collapse in a fortnight or disappear in a generation or even a lifetime, its claim to the future has passed. That is not to say that the Industrial Age will not remain with us. It will, just as other great economic epochs still do. One can still travel the backwaters of the planet and stumble upon faltering pockets of neolithic and even paleolithic life.

The industrial epoch marks the final stage of the age of fire. After thousands of years of putting fire to ore, the age of pyrotechnology is slowly burning out. Fire conditioned humankind's entire existence. In the *Protagoras,* Plato recounts how human beings came to possess fire and the pyrotechnological arts. According to the myth, as the gods began the process of fashioning living creatures out of earth and fire, Epimetheus and Prometheus were called upon to provide them with their proper qualities. By the time they came to human beings, Prometheus noticed that Epimetheus had already distributed all the qualities at their disposal to the rest of the plants and animals. Not wanting to leave human beings totally bereft, Prometheus stole the mechanical arts and fire from the gods and gave them to men and women. With these acquisitions, humanity acquired knowledge that originally belonged only to the gods.

Fire, said Lewis Mumford, provided human beings with light, power, and heat—three basic necessities for survival. Commenting on the role of fire in human development, Mumford concludes that its use "counts as man's unique technological achievement: unparalleled in any other species."[2] With fire, human beings could melt down the inanimate world of nature and reshape it into a world of pure utility. As the late historian Theodore Wertime of the Smithsonian Institution observed:

> There is almost nothing that is not brought to a finished state by means of fire. Fire takes this or that sand, and melts it, according to the locality, into glass, silver, cinnabar, lead of one kind or another, pigments or drugs. It is fire that smelts ore into copper, fire that produces iron and also tempers it, fire that purifies gold, fire that burns stone which causes the blocks in buildings to cohere [cement].[3]

The age of pyrotechnology began in earnest around 3000 B.C. in the Mediterranean and Near East when people shifted from the exclusive use of muscle power to shape inanimate nature to the use of fire. Pounding, squeezing, breaking, mashing, and grinding began to give way to fusing, melting, soldering, forging, and burning. By refiring the cold remains of what was once a fireball itself, human beings began the process of recycling the crust of the planet into a new home for themselves.

Now that humanity has fashioned this second home, it finds itself in short supply of the fossil fuels necessary to keep the economic furnaces afire while increasingly vulnerable to the effects of accumulating global-warming gases that threaten to radically change the Earth's climate. The industrial way of life has also become increasingly inhospitable to the rest of the Earth's creatures, who are largely unable to adjust to this alien man-made environment. Overpopulation, logging, grazing, and land development are resulting in massive deforestation and spreading desertification. The process is leading to the extinction of many of the remaining species of life, threatening a wholesale diminution of the Earth's biological diversity upon which we rely for sources of food, fiber, and pharmaceuticals. To put the magnitude of the problem in perspective, it is estimated that during the dinosaur age, species became extinct at a rate of about one per thousand years. By the early stages of the Industrial Age, species were dying out on the average of one per decade. Today, we are losing three species to extinction every hour.[4]

Humankind, then, faces three crises simultaneously—a dwindling of the Earth's nonrenewable energy reserves, a dangerous buildup of global-warming gases, and a steady decline in biological diversity. It is at this critical juncture that a revolutionary approach to organizing the planet is being advanced, an approach so far-reaching in scope that it will fundamentally alter humanity's relationship to the globe.

## A New Operational Matrix

Great economic changes in history occur when a number of technological and social forces come together to create a new "operating matrix." There are seven strands that make up the operational matrix of the Biotech Century. Together, they create a framework for a new economic era.

First, the ability to isolate, identify, and recombine genes is making the gene pool available, for the first time, as the primary raw resource for future economic activity. Recombinant DNA techniques and other biotechnologies allow scientists and biotech companies to locate, manipulate, and exploit genetic resources for specific economic ends.

Second, the awarding of patents on genes, cell lines, genetically engineered tissue, organs, and organisms, as well as the processes used to alter them, is giving the marketplace the commercial incentive to exploit the new resources.

Third, the globalization of commerce and trade make possible the wholesale reseeding of the Earth's biosphere with a laboratory-conceived second Genesis, an artificially produced bioindustrial nature designed to replace nature's own evolutionary scheme. A global life-science industry is already beginning to wield unprecedented power over the vast biological resources of the planet. Life-science fields ranging from agriculture to medicine are being consolidated under the umbrella of giant "life" companies in the emerging biotech marketplace.

Fourth, the mapping of the approximately 100,000 genes that comprise the human genome, new breakthroughs in genetic screening, including DNA chips, somatic gene therapy, and the imminent prospect of genetic engineering of human eggs, sperm, and embryonic cells, is paving the way for the wholesale alteration of the human species and the birth of a commercially driven eugenics civilization.

Fifth, a spate of new scientific studies on the genetic basis of human behavior and the new sociobiology that favors nature over nurture are providing a cultural context for the widespread acceptance of the new biotechnologies.

Sixth, the computer is providing the communication and organizational medium to manage the genetic information that makes up the biotech economy. All over the world, researchers are using computers to decipher, download, catalogue, and organize genetic information, creating a new store of genetic capital for use in the bioindustrial age. Computational technologies and genetic technologies are fusing together into a powerful new technological reality.

Seventh, a new cosmological narrative about evolution is beginning to challenge the neo-Darwinian citadel with a view of nature that is compatible with the operating assumptions of the new technologies and the new global economy. The new ideas about nature provide the legitimizing framework for the Biotech Century by suggesting that the new way we are reorganizing our economy and society are amplifications of nature's own principles and practices and, therefore, justifiable.

The Biotech Century brings with it a new resource base, a new set of transforming technologies, new forms of commercial protection to spur commerce, a global trading market to reseed the Earth with an artificial second Genesis, an emerging eugenics science, a new supporting sociology, a new communication tool to organize and manage economic activity at the genetic level, and a new cosmological narrative to accompany the journey. Together, genes, biotechnologies, life patents, the global life-science industry, human-gene screening and surgery, the new cultural currents, computers, and the revised theories of evolution are beginning to remake our world. All seven strands of the new operational matrix for the Biotech Century will be explored in the chapters that follow.

## Isolating and Recombining Genes

The new operational matrix began to take shape in the 1950s, when biologists discovered ways of locating and identifying chromosomes and genes. In the mid 1950s, cytologists—biologists who study the workings of cells—began experimenting with ways of separating chromosomes from the rest of a cell's makeup and organizing them so they could be analyzed under a microscope. The process is called karyotyping. In their book *Genome,* Jerry Bishop and Michael Waldholz point out the significance of the new cytogenetic tools: "For the first time, geneticists could correlate abnormalities in human chromosomes with genetic disease." The result was the birth of "a new science of medical genetics, embracing the study of human genetic disease at both the patient and chromosome levels."[5]

In 1968, Dr. Torbjorn O. Caspersson and Dr. Lore Zech, both cytochemists at the Karolinska Institute in Sweden, invented a process for identifying chromosomes, opening the door to the mapping of genes. The researchers realized that genes have different ratios of each of the four base nucleotides G, A, T, and C. They then found a chemical, acridine quinacrine mustard, that has an affinity for the G base, and stained the chromosome with it. When placed under ultraviolet light, the chromosome "glowed in a pattern of bright and dim spots reflecting high and low concentrations of base G."[6] Using the new banding pattern technique, Caspersson was able to identify individual human chromosomes. Other stains were subsequently developed and by the mid-1970s, researchers were studying alterations in banding patterns of chromosomes and connecting them to specific genetic traits and genetic disorders.

The first international workshop on gene mapping was convened in January of 1973 at Yale University in New Haven, Connecticut. Researchers reported on fifty newly mapped genes. At that time, only 150 genes had been mapped to specific chromosomes. By 1986, however, more than 1,500 genes had been mapped to specific chromosomes.[7] A year later, in 1987, Collaborative Research Inc., a small biotech start-up company in Bedford, Massachusetts, and researchers at MIT's Whitehead Institute, announced the compilation of "the first human genetic map."[8]

That same year, the United States Department of Energy (DOE) proposed an ambitious government-funded project to determine the sequence of all three billion G, A, T, and C base pairs that make up the human genome. Shortly thereafter, the National Institutes of Health (NIH) expressed its own interest in mapping the human genome and set up an Office of Human Genome Research to oversee the effort. In the fall of 1988, the government agencies agreed to join forces in the multibillion-dollar Human Genome Project, with NIH concentrating on gene mapping and the DOE on gene sequencing.[9] Other governments established their own human genome projects, followed closely on the heels by private commercial ventures. Genome projects have also been established for plants, microorganisms, and animal species.[10]

Currently, hundreds of millions of dollars are being spent on research all over the world to locate, tag, and identify the genes and the functions they serve in creatures throughout the biological kingdom. Vast amounts of genetic data on plants, animals, and human beings are being collected and stored in genetic databanks to be used as the primary raw resource for the coming Biotech Century.

The new methods for isolating, identifying, and storing genes are being accompanied by a host of new techniques to manipulate and transform genes. The most formidable of the new tools is recombinant DNA. In 1973, biologists Stanley Cohen of Stanford University and Herbert Boyer of the University of California performed a feat in the world of living matter that some biotech analysts believe rivals the importance of harnessing fire. The two researchers reported taking two unrelated organisms that could not mate in nature, isolating a piece of DNA from each, and then recombining the two pieces of genetic material. A product of nearly thirty years of investigation, climaxed by a series of rapid discoveries in the late

1960s and 1970s, recombinant DNA is a kind of biological sewing machine that can be used to stitch together the genetic fabric of unrelated organisms.

Cohen divides recombinant DNA surgery into several stages. To begin with, a chemical scalpel, called a restriction enzyme, is used to split apart the DNA molecules from one source—a human, for example. Once the DNA has been cut into pieces, a small segment of genetic material—a gene, perhaps, or a few genes in length—is separated out. Next, the restriction enzyme is used to slice out a segment from the body of a plasmid, a short length of DNA found in bacteria. Both the piece of human DNA and the body of the plasmid develop "sticky ends" as a result of the slicing process. The ends of both segments of DNA are then hooked together, forming a genetic whole composed of material from the two original sources. Finally, the modified plasmid is used as a vector, or vehicle, to move the DNA into a host cell, usually a bacterium. Absorbing the plasmid, the bacterium proceeds to duplicate it endlessly, producing identical copies of the new chimera. This is called cloned DNA.

The recombinant DNA process is the most dramatic technological tool to date in the growing biotechnological arsenal. The new techniques for identifying and manipulating genes are the first strand in the new operational matrix of the Biotech Century. After thousands of years of fusing, melting, soldering, forging, and burning inanimate matter to create useful things, we are now splicing, recombining, inserting, and stitching living material into economic utilities. Lord Ritchie-Calder, the British science writer, cast the biological revolution in the proper historical perspective when he observed that "just as we have manipulated plastics and metals, we are now manufacturing living materials."[11] We are moving from the age of pyrotechnology to the age of biotechnology. The speed of the discoveries is truly phenomenal. It is estimated that biological knowledge is currently doubling every five years, and in the field of genetics, the quantity of information is doubling every twenty-four months. The commercial possibilities, say the scientists, are limited only by the span of the human imagination and the whims and caprices of the marketplace.

Efficiency and speed lie at the heart of the new genetic engineering revolution. Nature's production and recycling schedules are deemed inadequate to ensure an improved standard of living for a burgeoning human population. To compensate for nature's slower pace, new ways must be found to engineer the genetic blueprints of microbes, plants, and animals in order to accelerate their transformation into useful economic products. Engineer the genetic blueprint of a tree so that it will grow to maturity quicker. Manipulate the genetic instructions of domestic breeds to produce faster-growing "super animals." Redesign the genetic information of cereal plants to increase their yield. According to a study by the United States government's now-defunct Office of Technology Assessment, bioengineering "can play a major role in improving the speed, efficiency, and productivity of . . . biological systems."[12] Our ultimate goal is to rival the growth curve of the Industrial Age by producing living material at a pace far exceeding nature's own time frame and then converting that living material into an economic cornucopia.

Some students of history might argue that human beings have been interested in increasing the quality and speed of production of biological resources since we first embarked on our agricultural way of life in the early neolithic era. That being the case, it might well be asked if genetic engineering is not simply a change in degree, rather than in kind, in the way we go about conceptualizing and organizing our relationship with the biological world. While the motivation behind genetic engineering is age-old, the technology itself represents something qualitatively new. To understand why this is the case, we must appreciate the distinction between traditional tinkering with biological organisms and genetic engineering.

We have been domesticating, breeding, and hybridizing animals and plants for more than ten millennia. But in the long history of such practices we have been restrained in what we could accomplish because of the natural constraints imposed by species borders. Although nature has, on occasion, allowed us to cross species boundaries, the incursions have always been very narrowly proscribed. Animal hybrids (mules, for example) are usually sterile, and plant hybrids do not breed true. As famed horticulturist Luther Burbank and a long line of his predecessors have understood, there are built-in limits as to how much can be manipulated when working at the organism or species level.

Genetic engineering bypasses species restraints altogether. With this new technology, manipulation occurs not at the species level but at the genetic level. The working unit is no longer the organism, but rather the gene. The implications are enormous and far-reaching.

To begin with, the entire notion of a species as a separate recognizable entity with a unique nature becomes an anachronism once we begin recombining genetic traits across natural mating boundaries. Three examples illustrate the dramatic change that genetic engineering makes in our relationship to nature.

In 1983, Ralph Brinster of the University of Pennsylvania Veterinary School inserted human growth hormone genes into mouse embryos. The mice expressed the human genes and grew twice as fast and nearly twice as big as any other mice. These "super mice," as they were dubbed by the press, then passed the human growth hormone gene onto their offspring. A strain of mice now exists that continues to express human growth genes, generation after generation. The human genes have been permanently incorporated into the genetic makeup of these animals.

Early in 1984, a comparable feat was accomplished in England. Scientists fused together embryo cells from a goat and from a sheep, and placed the fused embryo into a surrogate animal who gave birth to a sheep-goat chimera, the first such example of the "blending" of two completely unrelated animal species in human history.

In 1986, scientists took the gene whose product emits light in a firefly and inserted it into the genetic code of a tobacco plant. The tobacco leaves glow.

These results could never have been achieved even with the most sophisticated conventional breeding techniques. In the biotech labs, however, the recombinant possibilities are near limitless. The new genetic technologies allow us to combine genetic material across natural boundaries, reducing all of life to manipulatable chemical materials. This radical new form of biological manipulation changes both our concept of nature and our relationship to it. We begin to view life from the perspective of a chemist. The organism and the species no longer commands our attention or respect. Our interest now focuses increasingly on the thousands of chemical strands of genetic information that comprise the blueprints for living things.

With the newfound ability to identify, store, and manipulate the very chemical blueprints of living organisms, we assume a new role in the natural scheme of things. For the first time in history we become the engineers of life itself. We begin to reprogram the genetic codes of living things to suit our own cultural and economic needs and desires. We take on the task of creating a second Genesis, this time a synthetic one geared to the requisites of efficiency and productivity.

## Remaking the World

Today, hundreds of new bioengineering firms are setting the pace for the biotechnical revolution. With names like Amgen, Organogenesis, Genzyme, Calgene, Mycogen, and Myriad, these pioneers are blazing a trail for what some industry experts regard as the second great technological revolution in world history. Dozens of the world's leading transnational corporations are also pouring funds into biotechnical research. They include Du Pont, Novartis, Upjohn, Monsanto, Eli Lilly, Rohm and Haas, and Dow Chemical.

In nearly every life science field, development guidelines are being laid out, long-range retooling of equipment is being hurried along, new personnel are being hired, all in a mad rush to introduce the new genetic commerce into the economy, readying civilization to taste the first fruits of the biotechnological age.

There are already 1,300 biotech companies in the United States alone, with a total of nearly $13 billion in annual revenue, and more than 100,000 employees.[13] All of this development has occurred in only the first decade of a technological and economic revolution that will likely span several centuries. The Nobel Prize–winning chemist Robert F. Curl, of Rice University, spoke for many of his colleagues in science when he proclaimed that the twentieth century was "the century of physics and chemistry. But it is clear that the next century will be the century of biology."[14] The new biotechnologies are already reshaping virtually every field.

In the mining industry, researchers are developing new microorganisms that can replace the miner and his machine in the extraction of ores. As early as the 1980s, tests were being conducted with organisms that consume metals like cobalt, iron, nickel, and manganese. One company reported that it had successfully blown a certain bacterium "into low-grade copper ores where [it] produced an enzyme that eats away salts in the ore, leaving behind an almost pure form of copper."[15] For low-grade ores that are difficult to extract with conventional mining techniques, microorganisms will provide a more economical approach to extraction and processing. For example, scientists are now using microbial agents to degrade the minerals in which gold is trapped prior to chemical extraction, to increase the recovery rate of the gold.[16] In the future, the mining industry is expected to turn increasingly to bioleaching with microorganisms as a more economical way to utilize low-grade ores and mineral spoils that might ordinarily be discarded. Research is even going on to design microorganisms that can consume methane gas in the mines, eliminating one of the major sources of mine explosions.[17]

Energy companies are beginning to experiment with renewable resources as a substitute for coal, oil, and natural gas. Scientists hope to improve on existing crops, like sugar cane, which is already producing fuel for automobiles. Ethanol, derived from sugar and grain crops, is expected to provide more than 25 percent of U.S. motor vehicle fuel by the mid years of the coming century. Researchers are working on even more sophisticated approaches to biofuels, in the hope of replacing fossil fuels altogether. Scientists recently developed a strain of *E. coli* bacteria that can consume agricultural residues, yard trimmings, municipal solid waste and paper sludge, converting it to ethanol.[18]

Scientists in the chemical industry are talking about replacing petroleum, which for years has been the primary raw material for the production of plastics, with renewable resources produced by microorganisms and plants. A British firm, ICI, has developed strains of bacteria capable of producing plastics with a range of properties, including variant degrees of elasticity. The plastic is one hundred percent biodegradable and can be used in much the same way as petrochemical-based plastic resins. In 1993, Dr. Chris Sommerville, the director of plant biology at the Carnegie Institution of Washington, inserted a plastic-making gene into a mustard plant. The gene transforms the plant into a plastics factory. Monsanto hopes to have the plastic-producing plant on the market by the year 2003.[19]

Researchers are experimenting with even more exotic ways of creating new fibers and packaging materials. The United States Army is inserting genes into bacteria that are similar to the genes used by orb-weaving spiders to make silk. The spiders' silk is among the strongest fibers known to exist. Scientists hope to grow the silk-gene-producing bacteria in industrial vats and harvest it for use in products ranging from aircraft parts to bulletproof vests.[20]

Biotechnology is also being looked to as a key tool in environmental cleanup. Bioremediation is the use of living organisms—primarily microorganisms—to remove or render harmless dangerous pollutants and hazardous waste. A new generation of genetically engineered organisms is being developed to convert toxic materials into benign substances. Researchers are using genetically engineered fungi, bacteria, and algae as "biosorption" systems to capture polluting metals and radionuclides including mercury, copper, cadmium, uranium, and cobalt.[21] One biotech company, The Institute for Genomic Research, has successfully sequenced a microbe that can absorb large amounts of radioactivity. Company scientists hope to use the genes that code for the "uranium-gobbling pathway" to fashion new biological means of cleaning up radioactive dump sites.[22] With more than 200 million tons of hazardous materials being generated annually in the U.S. alone, and the costs of cleaning up toxic waste sites now estimated to be in excess of $1.7 trillion, industry analysts see bioremediation as one of the growth industries in the Biotech Century.[23]

Forestry companies have also turned to the new science in hopes of finding genes that can be inserted into trees to make them faster growing, resistant to disease, and better able to withstand heat, cold, and drought. Scientists at Calgene recently isolated a gene for the enzyme that controls the formation of cellulose in plants. They hope to enhance the enzyme to create trees with more cellulose in their cell walls, making a more efficient tree for harvesting in the pulp and paper-making industry.[24]

In agriculture, bioengineering is being looked to as a partial substitute for petrochemical farming. Scientists are busy at work engineering new food crops that can take in nitrogen directly from the air, rather than having to rely on the more costly petrochemical-based fertilizers currently in use. There are also experiments under way to transfer desirable genetic characteristics from one species to another in order to improve the nutritional value of plants and increase their yield and performance. Scientists are experimenting with genes that confer resistance to herbicides, help ward off viruses and pests, and can adapt a plant to salty or dry terrains, all in an effort to upgrade and speed the flow of agricultural products to market.

The first commercially grown gene-spliced food crops were planted in 1996. More than three-quarters of Alabama's cotton crop was genetically engineered to kill insects. In 1997, farmers planted genetically engineered soy on more than 8 million acres and genetically engineered corn on more than 3.5 million acres in the United States.[25] The chemical and agribusiness companies hope to see a majority of farmlands converted over to gene-spliced crops within the next five years.

Meanwhile, the first genetically engineered insect, a predator mite, was released in Florida in 1996. Researchers at the University of Florida hope it will eat other mites that damage strawberries and other crops. Scientists at the University of California at Riverside are inserting a lethal gene into the pink bollworm, a caterpillar that causes millions of dollars of damage to the nation's cotton fields each year. The killer gene becomes activated in the offspring, killing young caterpillars before they can damage the cotton, mate, and reproduce. Researchers Thomas Miller and John Peloquin hope to raise millions of the genetically engineered bollworms to adulthood and then release them into the environment to mate with wild bollworm moths. The offspring will contain the lethal gene and die en masse in this new form of pest management.[26]

Several biotech companies are working in the new field of tissue culture research, with the goal of moving more agricultural production indoors in the coming century. In the late 1980s, a U.S.-based biotechnology firm, Escagenetics of San Carlos, California—now defunct—announced it had successfully produced vanilla from plant-cell cultures in the laboratory. Vanilla is the most popular flavor in America. One-third of all ice cream sold in America is vanilla. Vanilla is expensive to produce, however. The plant has to be hand-pollinated and requires special attention in the harvesting and curing processes. Now, the new gene-splicing technologies allow researchers to produce commercial volumes of vanilla in laboratory vats—by isolating the gene that encodes the metabolic pathway that yields vanilla flavor and growing it in a bacteria bath—eliminating the bean, the plant, the soil, the cultivation, the harvest, and the farmer.

Researchers have also successfully grown orange and lemon vesicles from tissue culture, and some industry analysts believe that the day is not far off when orange juice will be "grown" in vats, eliminating the need for planting orange groves.[27] Scientists at the U.S. Department of Agriculture (USDA) have "tricked" loose cotton cells into growing by immersing them in a vat of nutrients. Because the cotton is grown under sterile conditions, free of microbial contamination, researchers say it could be used to make sterile gauze.[28]

The late Martin H. Rogoff and Stephen L. Rawlins, both former biologists and research administrators with the USDA, envision a hybrid form of agriculture production in both the fields and the factory. Fields would be planted only with perennial biomass crops. The crops would be harvested and converted to sugar solution by the use of enzymes. The solution would then be piped to urban factories and used as a nutrient source to produce large quantities of pulp from tissue cultures. The pulp would then be reconstituted and fabricated into different shapes and textures to mimic the traditional forms associated with "soil-grown" crops. Rawlins says that the new factories would be highly automated and require few workers.[29]

The many changes taking place in agriculture are being accompanied by revolutionary changes in the field of animal husbandry. Researchers are developing genetically engineered "super animals" with enhanced characteristics for food production. They are also creating novel transgenic animals to serve as "chemical factories" to produce drugs and medicines and as organ "donors" for human transplants. At the University of Adelaide in Australia, scientists have developed a novel breed of genetically engineered pigs that are 30 percent more efficient and brought to market seven weeks earlier than normal pigs. The Australian Commonwealth Scientific and Industrial Organization has produced genetically engineered sheep that grow 30 percent faster than normal ones and are currently transplanting genes into sheep to make their wool grow faster.[30]

At the University of Wisconsin, scientists genetically altered brooding turkey hens to increase their productivity. Brooding hens lay one-quarter to one-third fewer eggs than nonbrooding hens. As brooding hens make up nearly 20 percent of an average flock, researchers were anxious to curtail the "brooding instinct" because "broodiness disrupts production and costs producers a lot of money." By blocking the gene that produces the prolactin hormone, biologists were able to limit the natural brooding instinct in hens. The new breed of genetically engineered hens no longer exhibits the mothering instinct. They do, however, produce more eggs.[31]

Much of the cutting edge research in animal husbandry is occurring in the new field of "pharming." Researchers are transforming herds and flocks into bio-factories to produce pharmaceutical products, medicines, and nutrients. In April of 1996, Genzyme Transgenics announced the birth of Grace, a transgenic goat carrying a gene that produces BR-96, a monoclonal antibody being developed and tested by Bristol-Myers Squibb to deliver conjugated anti-cancer drugs. By the time Grace is one year old, she is expected to produce more than a kilogram of the experimental anti-cancer drug. Genzyme is also preparing to test a transgenic goat who produces anti-thrombin, an anti-clotting drug. Companies like Genzyme hope to produce drugs at half the cost by using transgenic pharm animals as chemical factories in the coming years. The company's CEO makes the point that Genzyme's new $10 million facility, which makes drugs for Gaucher disease, could be replaced in the near future with a herd of just twelve goats. Grace, by the way, is worth $1 million, making her the most valuable goat in history.[32]

Not to be outdone, researchers at PPL Therapeutics, in Blacksburg, Virginia, announced the birth of a transgenic calf named Rosie in February of 1997. The cow's milk contains alpha-lactalbumin, a human protein that provides essential amino acids, making it nutritious for premature infants who cannot nurse.[33] In Boulder, Colorado, Somatogen has created transgenic pigs who produce human hemoglobin.[34]

The new pharming technology moved a step closer to commercial reality on February 22, 1997, when Ian Wilmut, a fifty-two-year-old Scottish embryologist, announced the cloning of the first mammal in history—a sheep named Dolly. Wilmut replaced the DNA in a normal sheep egg with the DNA from the mammary gland of an adult sheep. He tricked the egg into growing and inserted it into the womb of another sheep.[35] The birth of Dolly is a milestone event of the emerging Biotechnological Age. It is now possible to mass-produce identical copies of a mammal, each indistinguishable from the original.

Shortly after the announcement of Dolly's birth, Wilmut and a research team led by Dr. Keith Campbell of PPL Therapeutics reported the birth of a second cloned sheep named Polly who contains a customized human gene in her biological code. Researchers added a human gene to fetal sheep cells growing in a laboratory dish and then cloned a sheep from the enhanced cells. The experiment caught even normally staid scientists by surprise. "After Dolly, everyone would have expected this, but they were saying it would happen in five to ten years," said Dr. Lee Silver, a molecular geneticist at Princeton University.[36]

Together, genetic manipulation and cloning will allow scientists to both customize and mass-produce animals, using the kind of quantifiable standards of measurement, predictability, and efficiency, that have heretofore been used to transform inanimate matter and energy into material goods. Agribusiness, pharmaceutical, and chemical companies plan to mass-produce customized and cloned animals for use as chemical factories, to secrete a range of drugs and medicines. The meat industry is also interested in cloning. Being able to reproduce animals with exacting standards of lean-to-fat ratios and other features provides a form of strict quality control that has eluded the industry in the past.

Animal clones will also be used to harvest organs for human transplantation. Being able to mass-produce exact replicas of animals will assure the kind of bioindustrial quality control that will be necessary to make xenotransplants a major commercial business in the Biotech Century. Biotech companies like Nextran and Alexion are inserting human genes into the germ lines of animal embryos to make their organs more compatible with the human genome and less likely to be rejected. Nextran is already in Phase I clinical trials to test the efficacy of using transgenic pig livers outside the body to help treat patients with acute liver failure, while they wait for a suitable human donor. In the procedure, doctors pump blood from a vein in the patient's leg through the pig liver, which is kept in a container at the patient's bedside. The blood is then pumped back into the patient's body through the jugular vein. Nextran's CEO, Marvin Miller, estimates the commercial value of his transgenic pig livers to be as high as $18,000 apiece.[37] With more than 100,000 Americans dying each year because a human organ was not available in time, the commercial market for xenotransplants is likely to be hefty. Salomon Brothers, the Wall Street investment company, estimates that more than 450,000 people, worldwide, will take advantage of xenotransplants by the year 2010. The market value of the new organ industry is likely to exceed $6 billion by then.[38]

Marine biotechnology is also expected to reap large profits in the coming decade. Scientists at Johns Hopkins University have already successfully transplanted an "anti-freeze" gene from flounder fish into the genetic code of bass and trout so that the fish will be able to survive in colder waters and provide new commercial opportunities for fishermen in northern climates. The Hopkins research team also inserted a mammalian growth hormone gene into fertilized fish eggs, producing faster-growing and heavier fish.[39] Other researchers are experimenting with the creation of sterile salmon who will not have the suicidal urge to spawn, but rather remain in the open sea to be commercially harvested. During the long journey back upstream to their birthing place, salmon stop eating and lose body weight. Scientists hope to break the reproduction cycle, by shocking salmon eggs to produce a doubling of the chromosomes, which results in the production of sterile fish. Michigan State University scientists say that by breaking the spawning cycle of chinook salmon, they can produce salmon whose body weight will exceed seventy pounds, compared to less than eighteen pounds for a fish returning to spawn.[40]

Most marine biotech research is geared to engineer customized fish that can be mass-produced through cloning techniques and be reared in fish farms. With one out of every five fish sent to market today coming from a fish farm, scientists hope the so-called "blue revolution" in marine biotechnology will rival the "green revolution" in agriculture. "By the year 2020," says Malcolm Beveridge, an ecologist at the University of Stirling's Institute of Aquaculture in Scotland, "[aquaculture] will be bigger than fisheries."[41]

Millions of people are already using genetically engineered drugs and medicines to treat heart disease, cancer, AIDS, and strokes. In 1995 researchers tested more than 284 new gene-spliced medicines, an increase of 20 percent over the previous year. Many conventional drugs have been replaced altogether with the new gene-spliced substitutes. Genetically engineered human insulin has virtually eliminated the use of naturally derived insulin from cows and pigs for more than 3.4 million Americans suffering from diabetes. Erythropoietin, produced by Amgen, is used by nearly 200,000 people who are on kidney dialysis each year. The gene-spliced product stimulates the growth of red blood cells, reducing the need for risky blood transfusions. Genentech's tissue plasminogen activator (tPA) dissolves blood clots. Avonex and Betaseron, the beta-interferons, are used as therapies for multiple sclerosis. Pulmozyme (DNase) is used to treat lung congestion in cystic fibrosis patients.[42]

These new genetically engineered drugs are only the beginning of the vast possibilities that lie ahead, say researchers in the field. Scientists in a number of genetic engineering laboratories are working on new ways of altering the genetic characteristics of insects who are the carriers of deadly human diseases, rendering them harmless as infective agents. Researchers at the National Institute of Allergy and Infectious Diseases are genetically engineering mosquitoes to make them unable to spread serious diseases. In one set of experiments, scientists have genetically engineered mosquitos with altered salivary glands making the insect unable to inject malarial parasites when it bites its victims. At Yale University, a medical research team is introducing "disease-prevention" genes into bacteria that live in the intestine of an insect called the "kissing bug." The bug, which is native to South America, spreads a parasite that causes the deadly Chagas disease. The genetically altered bacteria produce an antibiotic that kills the disease-causing parasite in the insect's digestive tract.[43]

Scientists claim that some of their current research with animal models may offer new hope for cures for diseases that have long been untreatable. In May of 1997, Drs. Se-Jin Lee, Ann M. Lawler, and Alexandra McPherron of Johns Hopkins University reported on their discovery of a gene that regulates growth in the muscle cells of mice. The researchers, who are affiliated with a Baltimore biotech company called MetaMorphix, found that the isolated gene produces a protein, myostatin, that regulates and controls muscle growth. When the regulatory gene is deleted, the mice grow more muscle. The researchers deleted the muscle-regulating gene in mouse embryo cells. The first of the new breed of mice, dubbed "Mighty Mouse," developed bulging muscles, huge shoulders, and broad hips. MetaMorphix hopes the research will lead to promising new treatments for muscle-related diseases like muscular dystrophy, and for diseases resulting in the wasting away of muscle like AIDS and cancer.[44]

Even more astounding, a Japanese research team reported in May of 1997 that they had successfully transplanted an entire human chromosome into the genetic code of mice, a feat thought unattainable by most scientists in the field. The Japanese team, led by Kazuma Tomizuka of the Kirin Brewery Technology Laboratory in Yokohama, fused human skin cells containing the chromosomes into mouse embryo cells. Some of the mouse embryo cells took up human chromosomes 14 and 22, which contain the genes that make human antibodies. The researchers took those embryonic cells and implanted them into female mice. The offspring carried the human chromosomes and produced antibodies made of human components when a foreign protein was introduced into their bodies.

Although scientists had inserted DNA into animals for years, they were only able to transplant a small bit at a time. The human chromosome inserted into mice in Japan contained nearly a thousand genes, or fifty times the amount previously transferred. Howard Petrie, director of the monoclonal antibody core facility at Memorial Sloan-Kettering Cancer Center, called the experiments "stunning." Petrie said the Japanese breakthrough meant that animals in the future might be engineered and mass-produced through clonal propagation and be used to make virtually unlimited amounts of therapeutic products such as human antibodies. In elevated doses, the antibodies could shrink tumors and kill viruses and bacteria.[45]

## Transforming Ourselves

Some biotech companies have been concentrating their efforts in the new field of tissue engineering and the fabrication of human organs. Hospitals are now using artificial skin, cultured and grown in the laboratory, to treat serious burn victims around the country. Recently a sixteen-year-old boy suffering from severe burns over 60 percent of his body was admitted to the University of California at San Diego Hospital. Attending physicians in the burn unit covered his wounds with skin that had been cultured and grown at Advanced Tissue Sciences. The boy was treated and released forty-seven days later.[46] Biotech companies like Advanced Tissue Sciences in La Jolla, California, and Genzyme Tissue Repair in Boston, Massachusetts, are pioneering the new field.

Researchers hope to move beyond the notion of transplants and into the era of fabrication, and are already well along in research to fabricate human heart valves, breasts, ears, cartilage, noses, and other body parts. "The idea is to make organs, rather than simply to move them," say Robert Langer and Dr. Joseph P. Vacanti, the two men most responsible for advancing the new field.[47] Langer and Vacanti first teamed up in 1984, when

Vacanti was a surgeon at Harvard Medical School and Langer was a chemical engineer at M.I.T. The theory behind the new technology is relatively simple. Here's how Langer and Vacanti explain the process.

> Using computer-aided design and manufacturing methods, researchers will shape the plastics into intricate scaffolding beds that mimic the structures of specific tissues and even organs. The scaffolds will be treated with compounds that help cells adhere and multiply, then "seeded" with cells. As the cells divide and assemble, the plastic degrades. Finally, only coherent tissue remains. The new, permanent tissue will then be implanted into the patient.[48]

A Boston company, Organogenesis, boasts that it can take a few cells from a human foreskin and "manufacture four acres of skin."[49] Companies like Organogenesis are demonstrating that a functional organ can grow from a few cells on a polymer frame. Cells, it turns out, are very adept at organizing the regeneration of their own tissue, and, according to Langer and Vacanti, "are able to communicate in three-dimensional culture using the same extracellular signals that guide the development of organs in utero."[50]

David Mooney, a chemical engineer at the University of Michigan, and Dr. James Martin of Carolina Medical Center are working on growing women's breasts in the laboratory. They hope, soon, to seed a three-dimensional scaffold molded in the shape of a breast with breast cells and then implant it into a woman's chest. The fabricated cells would then grow on the scaffolding until a new, living breast was formed.

Researchers around the country are also experimenting with the creation of fabricated lungs, hearts, livers, and pancreases made of human cells. At Boston's Children's Hospital, Dr. Anthony Atala, the director of tissue engineering at Harvard Medical School, is growing a human bladder in a glass jar. Atala's research team seeded a plastic scaffolding made to represent the three-dimensional shape of a bladder with bladder cells from a ten-year-old male patient and let the cells grow over the frame in a laboratory jar. Atala has been given approval to begin human trials and expects to transplant this laboratory-grown human organ into his young patient in 1998, making it the first tissue-engineered organ ever transplanted into a human being. Eventually, the scaffolding over which the cells have been growing will be destroyed by the patient's own enzymes, leaving a fully functioning human bladder in the boy's body. Atala's team is also working on growing human kidneys in laboratory jars. Researchers in the new field predict that by the year 2020, 95 percent of human body parts will be replaceable with laboratory-grown organs.[51]

Langer and Vacanti say that engineered structured tissue will replace plastic and metal prostheses for bones and joints in the coming century. "These living implants," say the researchers, "will merge seamlessly with the surrounding tissue, eliminating problems such as infection and loosening at the joint that plague contemporary prostheses."[52] Researchers have already customized noses and ears from polymer constructs and implanted them in animals in the laboratory. They anticipate the day when more complex body parts like human hands and arms will be grown on polymer scaffolding, customized to the needs of individual patients. The last remaining obstacle, caution the scientists, "is the resistance of nervous tissue to regeneration." No one has yet succeeded in growing human nerve cells, but most scientists working in the field are confident this remaining hurdle will be crossed in the very near future.[53]

Nowhere are the changes in molecular biology having a greater impact than in the fields of genetic screening and gene therapy. The Human Genome Project, the $3 billion government-sponsored program designed to map and sequence the entire human genome of approximately 100,000 genes by the year 2002, is redefining our notions of illness and our approaches to health care.[54]

Scientists hope to isolate and identify the gene or genes responsible for the more than four thousand genetic diseases that afflict human beings. They also hope to gain a better understanding of how genes function, turn "on" and "off," and interact with their environment to cause disease. Genetic screening tests are already available for many of the most common genetic diseases, and researchers expect that within less than a decade, an individual will be able to test for thousands of genetic diseases. Scientists are also researching the more complex polygenic disorders—involving clusters of genes—that affect mood, behavior, and personality.

A revolutionary new technology, DNA chips, will allow doctors to scan an individual's genomic makeup and provide a detailed readout of his or her genetic predispositions—a kind of crystal ball assessment of one's future mental and physical health. DNA chips are made up of thousands of different pieces of DNA placed on a silicon chip. The chips tag genetic differences, giving doctors a road map for sorting out an individual's existing and potential illnesses.[55]

Screening tests currently exist for breast cancer, Huntington's disease, Down syndrome, fragile X disease, cystic fibrosis, Tay-Sachs, Gaucher's disease, and sickle-cell anemia, to name just a few. The potential commercial market for genetic screening is estimated to be in the tens of billions of dollars by the early years of the twenty-first century.

The mapping and sequencing of the human genome is being accompanied by the first gene therapy trials on human patients. Over the past six years scientists have inserted foreign genes into hundreds of patients in an attempt to correct a number of genetic disorders. Although a recent NIH report criticized the protocols, saying none had yet proven effective, scientists hope that increasing knowledge of the genome and its functioning will make gene therapy a commonplace occurrence in the coming biotech century. Gene therapy has already been used in an attempt to treat ADA deficiency, cancer, and Parkinson's disease. Thus far, all of the therapy has been on somatic cells—although a number of scientists are now looking to correct genetic disorders at the germ line stage. In somatic therapy, the genetic changes affect only the individual patient, whereas, in germ line intervention, the genes are transplanted into the sperm, egg, or embryonic cells—and the genetic changes are passed along to future generations, affecting the evolution of the entire human species.

In April of 1997, researchers at Case Western Reserve University Medical School in Cleveland, Ohio, announced the creation of the first artificial human chromosome, a development that could lead to the customized design of genetic traits in the sex cells, or in embryonic cells just after conception. This newest breakthrough may one day "allow doctors to alter people's genetic inheritance or cure diseases by slipping genetic 'cassettes' directly into cells."[56]

The researchers blended both natural DNA and synthetic DNA "made by a machine" in the laboratory. The synthetic DNA was made to mimic part of a human chromosome called a centromere, which is the primary structure responsible for chromosome replication. The team, led by Dr. John J. Harrington and Huntington F. Willard, then injected the DNA into cells growing in laboratory dishes. The new DNA "self-assembled" into chromosomes. Genes on the artificial chromosome continued to function in daughter cells for more than six months and after the parent cell had divided more than 240 times. A Cleveland-based biotech company, Athersys Inc., owns the rights to the technology. Willard says that within a year his company plans on creating a modular system of prefabricated chromosome parts, "each bearing different key genes, which could be pulled off a laboratory shelf, combined and inserted into a person's cells."[57]

What makes the human artificial chromosome so potentially valuable, both as a medical technology and commercial product, is that it brings with it the kind of predictability that has, in the past, escaped scientists working in the fledgling field of gene therapy. Until now, scientists have had to insert individual genes into a virus and then use the virus as a vector to insert the genes into the cell's chromosomes. With this "shotgun" method, scientists can never know which chromosome will get the added gene or where the gene will locate along the chromosome once it arrives. There is no way to target the gene to the precise location desired. With artificial chromosomes, the process is more akin to inserting an entire genetic cassette into the body. Each gene is already in place on its own chromosome, eliminating the random nature of existing gene therapy techniques. Artificial chromosomes open up unlimited possibilities for modifying genetic structure, both in somatic and germ line cells.

Customizing genetic changes into a child, either before conception in the sex cells, just after conception in the embryonic cells, or during fetal development, is likely to become a reality within the next ten years. New breakthroughs in reproductive technologies, including the freezing and long-term storage of sperm, eggs, and embryos; in vitro fertilization techniques; embryo transplantation; and surrogacy arrangements, are revolutionizing human reproduction and conception and making possible, to an ever-increasing degree, the artificial manipulation of the unborn.

The first successful artificial insemination took place in 1884 at the Jefferson Medical School in Philadelphia. A married woman was inseminated with the sperm of a medical student. The practice did not become widespread, however, until the 1970s, when cryopreservation of sperm made it possible to perform artificial insemination using stored sperm, eliminating the inconvenience of having to match ovulation cycles with sperm donations. Storing large numbers of sperm samples also allowed infertile couples to select donor characteristics to assure a closer match to an infertile husband's own physiological makeup. Today, donor insemination clinics offer a wide selection of sperm samples to choose from. Prospective clients can select for race, height, body type, eye color, intelligence, ethnic and religious background, and even national origin.[58]

*In vitro* fertilization is a more recent development. In 1978, Lesley and John Brown of Oldham, England, gave birth to the first child conceived in a test tube. The birth of Louise Brown shocked the public and signaled the beginning of a new era in human reproduction. R. G. Edwards and Patrick Steptoe, the two British doctors who performed the first successful test-tube baby experiment, removed a single egg from Lesley Brown's ovary and placed it in a dish. John Brown's sperm was added to the dish and fertilization occurred. After waiting for the fertilized egg to divide three times, the physicians placed it into Louise Brown's uterus. Today, *in vitro* fertilization (IVF) is a widespread practice throughout the world. There are tens of thousands of IVF children, and IVF practitioners predict that within a decade, IVF babies could number in the hundreds of thousands worldwide.[59]

Many IVF clinics have begun to use frozen embryos as a more convenient way to ensure reproduction. The first child to be born from a frozen embryo was Zoe Leyland, on March 28, 1984, in Melbourne, Australia. In the procedure, eggs are harvested, fertilized, frozen, and stored for future implantation. The practice eliminates the stress of continuous hormonal stimulation of ovaries for egg retrieval.[60]

The increasing use of *in vitro* fertilization techniques has spawned a flourishing commercial surrogacy industry in several countries. Surrogate mothers contract with their clients—usually at a cost of $10,000 or more for the nine-month gestation period. There are several kinds of surrogate arrangements. Most often, the surrogate's own egg is fertilized with the male client's sperm. A growing number of surrogates, however, are gestational mothers. A frozen embryo from the clients is implanted in the surrogate's womb, where it is gestated. The birth child contains only the genetic information of the clients. The womb is, in a sense, rented for nine months to produce another couple's baby. To date, more than four thousand children have been born—most in the United States—of surrogate arrangements.[61] Tinkering with sperm, eggs, and embryos outside the womb opens up a range of new reproductive opportunities.

Being able to shape the genetic destiny of a human being before birth is also being helped along by new developments in the creation of artificial wombs. "The womb," cautioned the late Joseph Fletcher, former professor of medical ethics at the University of Virginia School of Medicine, "is a dark and dangerous place, a hazardous environment. We should want our potential children to be where they can be watched and protected as much as possible."[62] Already, scientists have succeeded in shortening the time unborn children need to be nurtured in the womb from nine to less than six months. At the same time, an increasing number of children start their lives outside the human womb in petri dishes where, as embryonic cells, they divide and grow before implantation into their own or a surrogate mother's womb. Being able to grow a fetus in a totally artificial womb, from conception to birth, say many scientists working in the new field of fetal molecular biology, would assure a more predictable environment and make it much easier to make genetic corrections and modifications.

The late French Nobel Prize–winning biologist Jean Rostand believed that the creation of the artificial womb was inevitable.[63] Today, researchers in laboratories around the world are working to make Rostand's prediction a reality. Some believe that the ability to grow a child completely outside the mother's womb, from conception to birth, is less than a decade away. Writing in *Scientific American* in 1995, Langer and Vacanti say the main stumbling block to keeping unborn children alive outside the human womb is that their immature lungs are unable to breathe air. They suggest a sterile, fluid-filled artificial womb.

> The babies would breathe liquids called perfluorocarbons, which carry oxygen and carbon dioxide in high concentrations. . . . A pump would maintain continuous circulation of the fluid, allowing for gas exchange. . . . The womb would [also] be equipped with filtering devices to remove toxins from the liquid. Nutrition would be delivered intravenously, as it is now. The womb would provide a self-contained system in which development and growth could proceed normally until the baby's second "birth."[64]

At least one scientist says he may be able to grow headless human clones in artificial wombs sometime early in the next century, to be used as spare parts during the lifetime of the human donors whose cells have been cloned. In October of 1997, Dr. Jonathan Slack, a professor of developmental biology at Bath University in the United Kingdom, reported that he and his colleagues were able to manipulate certain genes in a frog embryo and suppress the development of the tadpole's head, trunk, and tail. The experiment resulted in the birth of a live frog without a head. Slack said the same genes perform similar functions in both frogs and humans, raising the prospect of growing human body parts in artificial glass wombs. In an interview with *The Sunday Times* of London, Slack said that it occurred to him that "instead of growing an intact embryo, you could genetically reprogram the embryo to suppress growth in all the parts of the body except the bits you want, plus a heart and blood circulation."[65] Slack added that while it would be morally wrong to gestate organs in a woman's womb, "more acceptable might be taking a single cell and somehow growing a complete organ in a bottle."[66]

By cloning one's organs, the donor would be assured of a perfect match in a future organ transplant and would not have to run the risk of tissue rejection. Because the cloned organs would not have a head or nervous system, it might be possible to bypass conventional legal restrictions and ethical concerns governing experimentation on human embryos and fetuses. Although some ethicists voiced concern over Dr. Slack's experiment and its implications, a number of biologists offered their encouragement. Dr. Lewis Wolpert, professor of biology at University College, London, said that Slack's research was sensible and added that "there are no ethical issues because you are not doing harm to anyone."[67]

While engineering genetic modifications into unborn children is likely to happen within the next decade or so, the ability to gestate human fetuses in artificial wombs is a more distant prospect but still likely to be achieved well before the mid decades of the Biotech Century. Alan Bernstein, director of the Samuel Lunenfeld Research Institute of Mr. Sinai Hospital in Toronto, speaks of the profound changes taking place in our concept of human nature as a result of the many breakthroughs occurring in genetic science.

> Soon we will have all the instructions on how to make a human being—what thinking means and what memory means. It will totally transform how we view ourselves and disease. We can't anticipate all the ways it will impact us.[68]

We are about to remake ourselves as well as the rest of nature with little preparation and even less discussion of where this journey might end. What is clear, however, is that our notions about life will likely be fundamentally changed in the coming biotech century.

## From Alchemy to Algeny

The very thought of recombining living material into an infinite number of new combinations is so extraordinary that the human mind is barely able to grasp the immensity of the transition at hand. These first few processes and products are the biotechnical equivalent of the first pots and bins forged by our ancestors thousands of years ago when they began experimenting with the pyrotechnical arts for the first time. From the moment our neolithic kin first fired up the earth's material, transforming it into new forms, humanity locked itself into a long journey that finally culminated in the Industrial Age. Now humanity has set its sights on the living world, determined to reshape it into new combinations, and the far-distant consequences of this new journey are as unfathomable to today's biotechnologists as the specter of industrial society would have been to the first pyrotechnologists.

This great biotechnological transformation is being accompanied by an equally significant philosophical transformation. Humanity is beginning to reshape its view of existence to coincide with its new organizational relationship with the earth. The best way to understand this conceptual revolution is by comparing two powerful metaphors. For most of the pyrotechnical age, alchemy served as both the philosophical framework and conceptual guide to human beings' technological manipulation of the natural world. As late as the eighteenth century, Sir Isaac Newton, one of the founders of modern science, was still practicing the alchemical arts. Today, the stage is being set for the emergence of a new kind of consciousness—one that reflects the aspirations and objectives of the new biotechnical arts. "Algeny" is likely to emerge as a new philosophical framework and overarching metaphor for the Biotech Century. The term was first coined by Dr. Joshua Lederberg, the Nobel Laureate biologist and past president of Rockefeller University. I subsequently refined the meaning of the term in the 1980s. Algeny means to change the essence of a living thing. The algenic arts are dedicated to the "improvement" of existing organisms and the design of wholly new ones with the intent of "perfecting" their performance. But algeny is much more. It is humanity's attempt to give metaphysical meaning to its emerging technological relationship with nature. Algeny is a way of thinking about nature, and it is this new way of thinking that sets the course for the next great epoch in history.

When we think of the term "alchemy" today, what immediately comes to mind is the futile search for a method by which lead could be transformed into gold. As Morris Berman points out in his book, *The Reenchantment of the World,* it was much more. Alchemy was, at one and the same time, "the science of matter, the attempt to unravel nature's secrets; a set of procedures which were employed in mining, dyeing, glass manufacture, and the preparation of medicines, and simultaneously a type of yoga, a science of psychic transformation."[69]

Alchemy is said to have originated as a formal philosophy and process in Egypt during the fourth century B.C., although many historians believe that its roots lie much further in antiquity, dating as far back as the first city-state at Sumer. According to the alchemists, "All metals are in the process of becoming gold."[70] They are, in other words, gold *in potentia.* The alchemists believed that every metal was continually seeking to transform itself, to transcend its original state and experience its true nature, which they believed was gold. Meister Eckhart, the thirteenth-century mystic, once remarked that "copper is restless until it becomes gold."[71] The alchemists were firmly convinced that it was possible to accelerate what they believed to be the "natural process" of transformation by way of an elaborately orchestrated set of laboratory procedures.

The alchemic process began with the fusing together of several metals into one mass or alloy, which was then regarded as a kind of universal base material from which the various transmutations could be made. Fire, of course, was indispensable to the entire transitional process. It allowed the alchemist to melt, fuse, purify and distill his base material, creating new combinations and forms, each one closer to the ideal golden state.

Alchemy, then, was a philosophy and a technical activity at the same time. Nature, for the alchemist, was a process attempting to complete itself. The alchemist, as Morris Berman points out, viewed himself as a midwife, "accelerating" the natural process, hurrying the physical world along to its own perfected state. Alchemy is believed to have derived from an Arabic word meaning "perfection." Humanity's task, according to the alchemists, was to help nature in its struggle to "perfect itself." Gold, in its seeming permanence, conjured up the image of immortality and perfection.

Anthropologists believe that the alchemic tradition grew out of the arts of fabric dyeing and the bronzing or coloring of metals, which reached its zenith in ancient Alexandria. The precursors of the alchemists were artisans attempting to find cheap metal substitutes for gold. What began as art imitating nature eventually became transformed by the late Middle Ages into an all-embracing explanation of the workings, purpose, and goal of the natural order itself. So convinced were the alchemists that what they were doing in the laboratory was an integral part of the natural process that they came to believe that the "gold" they created was not an imitation at all but rather a superior form of gold, one that represented the perfect state toward which all natural gold aspired.

Now, as we move from a pyrotechnical to a biotechnical relationship with nature, a new conceptual metaphor is emerging. Algeny is about to give definition and purpose to the age of biotechnology. An algenist views the living world as *in potentia.* In this regard, the algenist doesn't think of an organism as a discrete entity but rather as a temporary set of relationships existing in a fluid context, on the way to becoming something else. For the algenist, species boundaries are just convenient labels for identifying a familiar biological condition or relationship, but are in no way regarded as impenetrable walls separating various plants and animals. Dr. Thomas Eisner, professor of biology and director of the Institute for Research in Chemical Ecology at Cornell University in Ithaca, New York, asks us to rethink our very notion of "species." He writes,

> As a consequence of recent advances in genetic engineering, [a biological species] must be viewed . . . as a depository of genes that are potentially transferable. A species is not merely a hard-bound volume of the library of nature. It is also a loose-leaf book, whose individual pages, the genes, might be available for selective transfer and modification of other species.[72]

The algenist contends that all living things are reducible to a base biological material, DNA, which can be extracted, manipulated, recombined, and programmed into an infinite number of combinations by a series of elaborate laboratory procedures. By engineering biological material, the algenist can create "imitations" of existing biological organisms that to his mind are of a superior nature to the originals being copied.

The final goal of the algenist is to engineer the perfect organism. The "golden state" is the state of optimal efficiency. Nature is seen as a hierarchical order of increasingly efficient living systems. The algenist is the ultimate engineer. His task is to "accelerate" the natural process by programming new creations that he believes are more "efficient" than those that exist in the state of nature.

Algeny is both philosophy and process. It is a way of perceiving nature and a way of acting on nature at the same time. It is a revolution in thought commensurate in scale to the revolution in technology that is emerging. We are moving from the alchemic metaphor to the algenic metaphor.

The short-term benefits of this extraordinary new power are seductive. We are being deluged almost daily by a stream of reports from the scientific community, industry, and government telling of the great advances in store for society in the wake of the biotechnology revolution. With our newfound power to manipulate the genetic code of life, we open up a new vista of virtually unlimited possibilities. It is no wonder that the Biotechnical Age is being touted as a monumental leap forward for the human race. Scientists, corporate spokespersons, and political leaders are extolling the virtues of this new technology with such enthusiasm that even the doubters among us are apt to be swept up in the excitement of the moment.

Yet, if history has taught us anything, it is that every new technological revolution brings with it both benefits and costs. The more powerful the technology is at expropriating and controlling the forces of nature, the more exacting the price we will be forced to pay in terms of disruption and destruction wreaked on the ecosystems and social systems that sustain life. Certainly our recent experience with both the nuclear and petrochemical revolutions bears out this most ancient truth.

It is troubling, then, to hear our scientists, corporate leaders, and politicians talk in such unqualified terms of the great marvels that await us in the coming biotech century. One can only conclude that in their zeal they are either being unmindful of the lessons of history or are being disingenuous in their public pronouncements.

Genetic engineering represents the ultimate tool. It extends humanity's reach over the forces of nature as no other technology in history, perhaps with the one exception of the nuclear bomb, the ultimate expression of the age of pyrotechnology. With genetic technology we assume control over the hereditary blueprints of life itself. Can any reasonable person believe for a moment that such unprecedented power is without substantial risk?

# Two Patenting Life

Genes are the "green gold" of the biotech century. The economic and political forces that control the genetic resources of the planet will exercise tremendous power over the future world economy, just as in the industrial age access to and control over fossil fuels and valuable metals helped determine control over world markets. In the years ahead, the planet's shrinking gene pool is going to become a source of increasing monetary value. Multinational corporations and governments are already scouting the continents in search of the new "green gold," hoping to locate microbes, plants, animals, and humans with rare genetic traits that might have future market potential. Once having located the desired traits, biotech companies are modifying them and then seeking patent protection for their new "inventions." Patenting life is the second strand of the new operational matrix of the Biotech Century.

A battle of historic proportions has emerged between the high-technology nations of the North and the poor developing nations of the South over the ownership of the planet's genetic treasures. The struggle for control of genetic resources has dominated the political agenda at the United Nations Food and Agriculture Organization's meetings for more than a decade. Some third world leaders argue that multinational corporations and Northern Hemisphere nations are attempting to seize the biological commons, most of which is found in the biologically rich tropical regions of the Southern Hemisphere. The Southern Hemisphere nations contend that genetic resources are part of their national heritage, just as oil is for the Middle East, and they should be compensated for their use. The multinational corporations and Northern Hemisphere nations maintain that genes increase in market value only when manipulated and recombined using sophisticated gene splicing techniques, and therefore they have no obligation to compensate countries from which the genes are taken. Two examples, one from agriculture, the other from the pharmaceutical field, illustrate the vast potential as well as the bitter struggle that surrounds the enclosure of the genetic commons.

Years ago, scientists discovered a rare perennial strain of maize growing in a mountain forest of south central Mexico. Only a few thousand stalks of the perennial strain existed, in three tiny patches, and they were about to be bulldozed by farmers and loggers. The newly discovered strain was found to be resistant to leaf fungus, which had devastated the U.S. corn crop in 1970, costing farmers over $2 billion. The commercial value of the strain could total several billion dollars a year, according to geneticists and seed company experts.[1]

The rosy periwinkle, found in the tropical rain forest of Madagascar, offers another graphic example of the potential profits that lie in store with the enclosure of the genetic commons. Several years ago, researchers discovered that the rare periwinkle plant contained a unique genetic trait that could be used as a pharmaceutical to treat certain kinds of cancer. Eli Lilly, the pharmaceutical company that developed the drug, is making significant profits—$160 million in sales in 1993 alone—while Madagascar has not received so much as a penny of compensation for the expropriation of one of its natural resources.

Governments around the world have already set up gene storage facilities to preserve rare strains of plants whose genetic traits may prove commercially useful in the future. The U.S. National Seed Storage Laboratory at Fort Collins, Colorado, contains more than 400,000 seeds from all over the world. Many nations are also beginning to establish additional gene banks to store rare microorganisms and frozen animal embryos. In the years ahead, the commercial value of many of these rare strains of plants and breeds of animals will increase dramatically as the world market relies on genetic technologies to produce materials and products.[2]

## Enclosing the Last Frontier

The worldwide race to patent the gene pool of the planet is the culmination of a five-hundred-year odyssey to commercially enclose and privatize all of the great ecosystems that make up the Earth's biosphere. The history of "enclosures" is critical to understanding the potential long-term consequences of current efforts to enclose the world's gene pool.

The commodification of the global commons began in Tudor England in the 1500s with the enactment of the great "enclosure acts." The laws were designed to privatize the feudal commons, transforming the land from a shared trust to private real estate that could be bought and sold as individual units of property in the commercial marketplace. The catastrophic change in people's relationship to the land touched off a series of economic and social reforms that would remake society and reshape humanity's relationship to the natural world for the whole of the modern era.

Much of the economic life of medieval Europe centered around the village commons. Although feudal landlords owned the commons, they leased it to peasant farmers under various tenancy arrangements. In return for their right to cultivate the land, tenant farmers had to turn over a percentage of their harvest to their landlord or devote a comparable amount of time to working the landlord's fields. With the introduction of a moneyed economy in the late medieval period, peasant farmers were increasingly required to pay rent or taxes in return for the right to farm the land.

Medieval European agriculture was communally organized. Peasants pooled their individual holdings into open fields that were jointly cultivated. Common pastures were used to graze their animals. Life was spare, demanding, and unpredictable.

The village commons existed for more than six hundred years along the base of the feudal pyramid, under the watchful presence of the landlords, monarch, and pope. Then, beginning in the 1500s, new and powerful political and economic forces were unleashed, first in Tudor England, and later on the continent, which undermined and ultimately destroyed the communitarianism of village life that had bound humans to one another and the land for centuries.

"Enclosing" means "surrounding a piece of land with hedges, ditches, or other barriers to the free passage of men and animals."[3] Enclosure placed the land under private control, severing any right the community formerly had to use it. The enclosure movement was carried out by several means, including acts of Parliament, the agreement of all members of the village commune, and license by the king.

Some historians have called the enclosure movement "the revolution of the rich against the poor."[4] Between the sixteenth and nineteenth centuries, a series of political and legal acts were initiated in countries throughout Europe that enclosed publicly held land. In the process, millions of peasants were dislodged from their ancestral homes and forced to migrate into the new towns and cities where, if they were fortunate, they might secure subsistence employment.

England experienced two major waves of enclosure, the first, as noted, in the 1500s under the Tudor monarchy, the second in the late 1700s and early 1800s during King George III's reign.[5] In the earlier period, increased urban demand for food triggered an inflationary spiral which, in turn, increased the cost to landlords whose land rents had been fixed at pre-inflationary rates. At the same time, an expanding textile industry was clamoring for more wool, making sheep grazing an increasingly attractive and lucrative prospect. Those two forces conspired, creating an irresistible lure to enclose the land.

With the financial help of a new and wealthy bourgeois class of merchants and bankers, landlords began to buy up the common lands, turning them into pastureland for sheep. Enclosure fundamentally restructured the way people perceived their relationships with each other and the soil. English historian Gilbert Slater asks us to imagine "what a village cataclysm took place" when an act of Parliament was passed to enclose a village commons.[6] Commissioners descended on the village with account books in hand. They went door to door, plot to plot, assigning a monetary value to every property. The commissioners then rearranged the entire commons, cutting up the arable land and pastures into neat, orderly rectangles, each with a separate owner. All past relationships and mutual obligations were severed. Suddenly and arbitrarily, the customs and traditions of generations were declared null and void. Neighbors were no longer expected to help plow each other's fields or share in the grazing of draft animals on the common pasture.

Enclosure introduced a new concept of human relationships into European civilization that changed the basis of economic security and the perception of social life. Land was no longer something people belonged to, but rather a commodity people possessed. Land was reduced to a quantitative status and measured by its exchange value. So, too, with people. Relationships were reorganized. Neighbors became employees or contractors. Reciprocity was replaced with hourly wages. People sold their time and labor where they used to share their toil. Human beings began to view each other and everything around them in financial terms. Virtually everyone and everything became negotiable and could be purchased at an appropriate price. Max Weber, the German economist and sociologist, called this great restructuring of relationships the "disenchantment of the world"—life, land, existence, reduced to abstract quantifiable standards of measurement. The European enclosure movement set the stage for the modern age.

The enclosure of the European landmass and the conversion of the feudal commons to privately held real estate began a process of privatization of the land commons around the world. Today, virtually every square foot of landmass on the planet—with the exception of Antarctica which, by international agreement, has been partially preserved as a non-exploitable shared commons—is either under private commercial ownership or government control.

The enclosure of the Earth's landmass has been followed, in rapid succession, by the commercial enclosure of parts of the oceanic commons, the atmospheric commons, and, more recently, the electromagnetic spectrum commons. Today large swaths of the ocean—near coastal waters—are commercially leased, as is the air which has been converted into commercial air corridors, and the electromagnetic frequencies which governments lease to private companies for radio, telephone, television, and computer transmission.

Now, the most intimate commons of all is being enclosed and reduced to private commercial property that can be bought and sold on the global market. The international effort to convert the genetic blueprints of millions of years of evolution to privately held intellectual property represents both the completion of a half-millennium of commercial history and the closing of the last remaining frontier of the natural world.

The enclosure and privatization of the planet's genetic commons began in 1971 when an Indian microbiologist, Ananda Chakrabarty, at the time an employee of the General Electric Company (G.E.), applied to the U.S. Patents and Trademark Office (PTO) for a patent on a genetically engineered microorganism designed to consume oil spills on the oceans. The

PTO rejected the patent request, arguing that living things are not patentable under U.S. patent law. To underscore its contention, the PTO pointed out that in the few cases where patents had been extended to life forms—for asexually reproducing plants—it had taken a legislative act of Congress to create a special exception.

Chakrabarty and G.E. appealed the PTO decision to the Court of Customs and Patent Appeals, where, to the surprise of many observers, they won by a narrow three-to-two decision. The majority on the court argued that "the fact that microorganisms . . . are alive [is] without legal significance."[7] The justices went on to say that the patented microorganism was "more akin to inanimate chemical compositions such as reactants, reagents, and catalysts, than [it was] to horses and honeybees or raspberries and roses."[8] Clearly, had the first patent request been for a genetically engineered mouse or chimpanzee, it is highly unlikely, given the justices' remarks that the microbe appeared to be closer to a chemical than a horse, that a patent would ever have been granted.

After additional judicial wrangling, the landmark case was appealed, once again, by the Patent Office, this time to the U.S. Supreme Court. The PTO was joined in the case by the People's Business Commission—shortly thereafter renamed The Foundation on Economic Trends—which provided the main amicus curiae brief. The brief, written by Ted Howard, claimed that the case before the court went directly to the heart of the question of the intrinsic value and meaning of life. If the patent were upheld by the court, Howard argued, then "manufactured life—high and low—will have been categorized as less than life, as nothing but common chemicals."[9] The People's Business Commission predicted that a favorable court decision would open the door to the patenting of all forms of life in the future.

In 1980, by a slim margin of five to four, the justices ruled in favor of Chakrabarty, granting a patent on the first genetically engineered life form. Speaking for the majority, Chief Justice Warren Burger argued that "the relevant distinction was not between living and inanimate things," but whether or not Chakrabarty's microbe was a "human-made invention."[10] Speaking for the minority, the late Justice William Brennan countered that "it is the role of Congress, not this court, to broaden or narrow the reach of patent laws." Brennan cautioned that in the case of patenting life, Congressional guidance is even more important since "the composition sought to be patented uniquely implicated matters of public concern."[11] All of the justices believed that their decision was a narrowly construed one. Justice Burger, referring to "the gruesome parade of horribles" outlined in the People's Business Commission amicus, made clear his conviction that the decision rendered by the court was an orthodox interpretation of existing patent law and not meant to open up the larger social issues surrounding the genetic engineering of life.[12] These concerns, said the justices, should be left to Congress.

The court's action laid the all-important legal groundwork for the privatization and commodification of the genetic commons. In the aftermath of that historic decision, bioengineering technology shed its pristine academic garb and bounded into the marketplace, where it was heralded by many analysts as a scientific godsend, the long-awaited replacement for a dying industrial order. Nelson Sneider, an investment analyst for E. F. Hutton, a firm that had nurtured much of the initial interest in the emerging biotechnical field, said at the time, "We are sitting at the edge of a technological breakthrough that could be as important as . . . [the] discovery of fire."[13] So anxious was Wall Street to begin financing the biotechnical revolution that when the first privately held genetic engineering firm offered its stock to public investors, it set off a buying stampede within the investment community. On October 14, 1980, just months after the Supreme Court cleared the way for the commercial exploitation of life, Genentech offered over one million shares of stock at $35 per share. In the first twenty minutes of trading, the stock climbed to $89 a share. By the time the trading bell had rung in late afternoon, the fledgling biotechnology firm had raised $36 million and was valued at $532 million. The astounding thing was that Genentech had yet to introduce a single product into the marketplace. Commented one financial analyst, "I have been with [Merrill Lynch] twenty-two years [and] I have never seen anything like this."[14]

Corporate America understood the profound implications of the court decision. Genentech gushed, "The court has assured the country's technology future."[15] Chemical, pharmaceutical, agribusiness, and biotech start-up companies everywhere sped up their research and development work, mindful that the granting of patent protection meant the possibility of harnessing the genetic commons for vast commercial gain in the years ahead. Some observers, however, were not so enthused. Ethicist Leon Kass asked,

> What is the principled limit to this beginning extension of the domain of private ownership and dominion over living nature . . . ? The principle used in Chakrabarty says that there is nothing in the nature of a being, not even in the human patentor himself, that makes him immune to being patented.[16]

Here, for the first time, was a judiciary claiming that, for purposes of commerce, there is no longer any need to make distinctions between living beings and inanimate objects. Henceforth, a genetically engineered organism was to be regarded as an invention in the same way as computers or other machines are considered inventions. If Chakrabarty's microbe could be patented, why not any other form of life that has been, in any way, genetically engineered? What might it mean for subsequent generations to grow up in a world where they come to think of all of life as mere invention—where the boundaries between the sacred and the profane, and between intrinsic and utility value, have all but disappeared, reducing life itself to an objectified status, devoid of any unique or essential quality that might differentiate it from the strictly mechanical?

## Life as Invention

While the Supreme Court decision lent an air of legal legitimacy to the emerging biotech industry, a PTO decision seven years later, in 1987, opened up the floodgates for the wholesale commercial enclosure of the world's gene pool, signaling the beginning of a new economic era in world history. The PTO, in a complete about-face, reversed its earlier position and issued a ruling that all genetically engineered multicellular living organisms, including animals, are potentially patentable.[17] The ruling was breathtaking in scope. In a single regulatory stroke, the PTO had placed the global economy on a new course—one that would lead us out of the Industrial Age and into the Biotech Century. The expansiveness of the decree was not lost on the Patent Office. The Commissioner of Patents and Trademarks at the time, Donald J. Quigg, attempted to calm a shocked public by asserting that the decision covered every creature but human beings. The reason for excluding humans, said Quigg, was that the Thirteenth Amendment to the Constitution forbids human slavery. On the other hand, genetically altered human embryos and fetuses as well as human genes, cell lines, tissues, and organs are potentially patentable, leaving open the possibility of patenting all of the separate parts, if not the whole, of a human being.

At the very heart of the issue of patentability is the question of whether engineered genes, cells, tissues, organs, and whole organisms are truly human inventions or merely discoveries of nature that have been skillfully modified by human beings. In order to qualify as a patented invention, the inventor must prove that the object is novel, non-obvious, and useful—that is, that no one has ever made the object before, that the object is not something that is so obvious that someone might have thought of it using prior art, and that the object serves some useful purpose. Against this standard is an equally compelling qualification. Even if something is novel, non-obvious, and useful, if it is a discovery of nature, it is not an invention and therefore not patentable. For this reason, the chemical elements, while unique, non-obvious when first isolated, and very useful, were nonetheless not considered patentable, as they were discoveries of nature. This is so despite the fact that some degree of human ingenuity went into isolating and classifying their properties.

What makes the Supreme Court decision and subsequent PTO ruling so audacious, and suspect from a legal point of view, is that it appears to defy the very logic of previous patent rulings that preclude claiming a discovery of nature as an invention. No molecular biologist has ever created a gene, cell, tissue, organ, or organism *de novo.* In this sense, the analogy between the elements of the periodic table and genes and living matter is appropriate. No reasonable person would dare suggest that a scientist who isolated, classified, and described the properties of hydrogen, helium, or oxygen ought to be granted the exclusive right, for twenty years, to claim the substance as a human invention. The PTO has, however, said that the isolation and classification of a gene's properties and purposes is sufficient to claim it as an invention.

The prevailing logic becomes even more strained when consideration turns to patenting a cell line, or a genetically modified organ, or a genetically modified whole animal. Is a pancreas or kidney patentable simply because it's been subjected to a slight genetic modification? What about a chimpanzee? Here's an animal who shares 99 percent of the genetic makeup of a human being and has the mental capacity of a two-year-old child. Should a chimpanzee qualify as a human invention if researchers insert a single gene into his or her biological makeup? The answer, according to the PTO, is yes.

William Tucker, manager for technology transfer at DNA Plant Technology in Oakland, California, says, "Just because it's biological and self-reproducing doesn't, to me, make it any different from a piece of machinery that you manufacture from nuts and bolts and screws."[18] Tucker, and other champions of life patents, forget that none of the genes, cells, tissues, organs, or organisms they are seeking patents on have been assembled and manufactured. In what sense is the breast cancer gene a human invention? or stem cells? or genetically engineered onco-mice? None of them have been assembled or manufactured, but simply isolated, synthesized, and modified.

Key Dismukes, former Study Director for the Committee on Vision of the National Academy of Sciences, spoke clearly and forcefully to the misguided logic underlying both the Supreme Court decision on Chakrabarty and subsequent policy formulations by the PTO on the patenting of life.

> Let us get one thing straight: Ananda Chakrabarty did not create a new form of life; he merely intervened in the normal processes by which strains of bacteria exchange genetic information to produce a new strain with an altered metabolic pattern. "His" bacterium lives and reproduces itself under the forces that guide all cellular life. Recent advances in DNA techniques allow more direct biochemical manipulation of bacterial genes than Chakrabarty employed, but these too are only modulations of biological processes. We are incalculably far away from being able to create life *de novo,* and for that I'm profoundly grateful. The argument that the bacterium is Chakrabarty's handiwork and not nature's wildly exaggerates human power and displays the same hubris and ignorance of biology that has had such devastating impact on the ecology of our planet.[19]

A year after issuing the new policy guidelines extending patentability across the biological commons, the PTO granted a patent on the first mammal, a genetically engineered mouse containing human genes that predispose it to developing cancer. The so-called "onco-mouse" was the "invention" of Harvard biologist Philip Leder. The mouse, which has been licensed to Du Pont, is sold as a "research model" for studying cancer.[20] Several other genetically engineered animals have been patented since, and nearly two hundred genetically modified animals, including pigs, cows, and sheep, are awaiting patent approval in the U.S.[21]

Recently, the Scottish research team that cloned the now famous sheep Dolly applied for a broad patent which would give it exclusive right of ownership over all cloned mammals. The patent application included human clones as well. As the legal status of human clones has not yet been resolved in public policy, they might not enjoy the same exclusion as human beings who are protected from patents in various countries because of existing statutes forbidding slavery.

Some of the life patents being granted in the United States are so broad, they give individual companies a virtual monopoly over the use of whole species. The patent awarded to Philip Leder extends to any animal whose germ line is engineered to contain cancer-causing genes.[22] Similarly broad patents have been awarded to chemical and pharmaceutical companies for major food crops. Agracetus, formerly a subsidiary of W. R. Grace Company, now of Monsanto, received a patent that "covers all cotton seeds and plants which contain a recombinant gene construction (i.e., are genetically engineered)."[23] Other companies, worried that "the patent would give a single multinational corporation unprecedented control over the world's staple crops," filed objections with the PTO, forcing the agency to reexamine the claim and the patent.[24] Concerned that it might have overstepped its authority in granting the broad patent, the PTO rejected its own prior approval. The case is now winding its way through the patent appeals process. In the meantime, the Agracetus patent remains valid, pending the final outcome of the appeals.

Undaunted by its temporary setback, W. R. Grace sought and received a second broad patent from the European Patent Office, in 1994, on a technique it "developed" "which makes possible the insertion of genes into any soybean variety." Like the decision on the cotton, the awarding of the patent sent the rest of the business and scientific community into a tailspin. Dr. Geoffrey Hawtin, director general of the International Plant Genetic

Resources Institute, expressed the anger of many of his colleagues when he warned,

> The granting of patents covering all genetically engineered varieties of a species, irrespective of the genes concerned or how they were transferred, puts in the hands of a single inventor the possibility to control what we grow on our farms and in our gardens. At the stroke of a pen, the research of countless farmers and scientists has potentially been negated in a single, legal act of economic highjack.[25]

The fierce competition between chemical, pharmaceutical, agribusiness, and biotech companies for commercial patents on genes, organisms, and processes to manipulate them is unprecedented. Charges of patent infringements, unfair use of prior art, stealing of trade secrets, and pirating of research, have resulted in a record number of legal challenges at the European and U.S. patent offices and in the courts as companies jostle with each other to improve their market share and competitive position.

## Biopiracy

Corporate efforts to enclose and commodify the gene pool are meeting with strong resistance from a growing number of countries and nongovernmental organizations (NGOs) in the Southern Hemisphere, which are beginning to demand an equitable sharing of the fruits of the biotech revolution. While the technological expertise needed to manipulate the new "green gold" resides in scientific laboratories and corporate boardrooms in the North, most of the genetic resources that are essential to fuel the new revolution lie in the tropical ecosystems of the South. The battle between Northern multinational corporations and Southern countries for control over the global genetic commons is likely to be one of the pivotal economic and political struggles of the Biotech Century.

In many important respects, the charges and countercharges, accusations and denials around what some refer to as "biopiracy," or "biocolonialism," are not new. The history of the colonial struggle has been one of continual usurpation and exploitation of native biological riches for the advantage of home markets. The great explorations of the New World were as dedicated to the task of finding new biological sources for food, fiber, dye, and medicine as to discovering gold, silver, and other rare metals. European nations established plantation colonies throughout the New World. Native knowledge of agriculture and native labor were both exploited to grow new food staples for export onto the world markets. Cassava, sweet potatoes, peanuts, maize, beans, and squash were among the many new native cash crops that were worth their weight in gold in the developing commodities market.

Explorers, and later Catholic missionaries and embassy personnel, devoted a great deal of time to biological prospecting, with the hope of securing new biological treasures that could be transformed into lucrative commercial markets. Many colonial nations, eager to maintain exclusive control over their biological conquests, enacted stiff penalties on plant contraband, including the imposition of the death penalty for the theft of valuable plants. On occasion, the theft of native resources could affect the futures of entire empires. The pirating of rubber plants from Brazil to Southeast Asia at the turn of the last century gave the British a commercial advantage in the critical world rubber market and led to the collapse of American efforts to control rubber production in the Western Hemisphere.[26]

Today, the plant hunters are giving way to the gene prospectors. Corporate giants are financing expeditions across the Southern Hemisphere in search of unusual and rare genetic traits that might have some commercial value. The potential stakes are enormous. Consider just the value of new drugs. Nearly three quarters of all the plant-based prescription drugs in use today were derived from drugs used in indigenous medicine. For example, curare, an important surgical anesthetic and muscle relaxant, is derived from plant extracts used by Amazonian Indians to stun prey.[27]

Southern countries claim that what Northern companies call "discoveries" are really the pirating of the accumulated indigenous knowledge of native peoples and cultures. It is true that Northern companies add some value by engineering and modifying the genetic makeup of plants, or by isolating out, purifying, distilling, and mass producing through clonal propagation and other means, genes that code for specific proteins that are useful in foods, medicine, fiber, and dyes. Still, Southern countries argue that a slight genetic modification of a crop or herb in the laboratory is rather insignificant, especially when measured against the centuries of painstaking stewardship required to nurture and preserve organisms containing those very unique and valuable traits so coveted by the scientists in their research.

The debate over Northern scientific knowledge versus Southern indigenous knowledge came to the fore in the controversy surrounding W. R. Grace's effort to patent certain processes for using the neem tree, a native tree of India. The neem tree patents created a furor in India and around the world and became a lightning rod for NGOs that oppose corporate attempts to patent indigenous knowledge and native biological resources.

The neem tree is an Indian symbol and enjoys an almost mystical status in that country. Ancient Indian texts refer to it as "the blessed tree" and the tree that "cures all ailments." The tree has been used for centuries as a source for medicines and fuel and is grown in villages across the Indian subcontinent. The New Year begins in India with the ritual eating of the tender shoots of the neem tree. Millions of Indians use neem twigs as toothbrushes to protect teeth because of the tree's antibacterial properties. The leaves and bark are used to cure acne and to treat a range of illnesses from infections to diabetes.[28] The tree has proven particularly valuable as a natural pesticide and has been used by villagers to protect against crop pests for hundreds of years. In fact, this natural pesticide is more potent that many conventional industrial insecticides. The neem extract wards off more than two hundred common insects that harm crops, including locusts, nematodes, boll weevils, beetles, and hoppers. A number of studies have found the neem extract to be at least as effective as synthetic insecticides like Malathion, DDT, and dieldrin, but without the deleterious impacts on the environment.[29]

W. R. Grace isolated the most potent ingredient in the neem seed, azadirachtin, and then sought and received a number of process patents for the production of the neem extract from the PTO. The company argued that the processes it used to isolate and stabilize the azadirachtin were unique and non-obvious. Indian scientists, however, in affidavits to the PTO, made clear that Indian researchers and companies had been treating neem seeds with the same processes and solvents as W. R. Grace, years before the company sought a patent on the processes. Moreover, Indian researchers and companies had never sought patent protection because they consider the information about the uses of the neem tree to be the result of centuries of indigenous research and development and something to be shared openly and freely. Now they are concerned that W. R. Grace's patents will deprive local farmers of their ability to produce and use neem-based pesticides by altering the price and availability of neem seeds. Outraged by W. R. Grace's usurpation of indigenous knowledge, Dr. B. N. Dhawan, emeritus scientist at the Central Drug Institute of India, said,

> It is really unfortunate that the benefits of all this work should go to an individual or to a company. I sincerely hope that the opinion of the international community will prevail and that the Neem [tree] will continue to be available for use by people all over the world without paying a high price to a company.[30]

Skirmishes over the perceived usurpation of indigenous knowledge and native resources are occurring with great frequency, as the global market makes its historic shift from an economy based on fossil fuels and rare metals to one increasingly based on genetic and biological resources.

Anxious to defuse the growing opposition, global companies are seeking to impose a uniform intellectual property regime—one that will be binding on every country and that will give the multinationals free access to genetic material from around the world while at the same time providing them protection for their genetically engineered products.

Global corporations went a long way toward achieving their ends with the passage of the Trade Related Aspects of Intellectual Property Agreement at the Uruguay Round of the General Agreement on Tariffs and Trade (GATT). The agreement, which was designed to create a uniform framework for intellectual property protection, was sculpted, in large part, by a coalition of companies calling themselves the Intellectual Property Committee (IPC). The participating firms include many of the major players in the biotech field including Bristol Myers, Merck, Pfizer, Monsanto, and Du Pont. Monsanto's James Enyart explained the rationale behind the IPC strategy.

> . . [O]ur Trilateral Group was able to distill from the laws of the more advanced countries the fundamental principles for protecting all forms of intellectual property. . . . Besides selling our concepts at home, we went to Geneva where [we] presented [our] document to the staff of the GATT

> Secretariat. We also took the opportunity to present it to the Geneva-based representatives of a large number of countries. . . . What I have described to you is absolutely unprecedented in GATT. Industry identified a major problem for international trade. It crafted a solution, reduced it to a concrete proposal, and sold it to our own and other governments . . . The industries and traders of world commerce have played simultaneously the role of patients, the diagnosticians, and the prescribing physicians.[31]

The TRIPS agreement makes no allowance for indigenous knowledge, notes Dr. Vandana Shiva, director of the Research Foundation for Science, Technology, and Natural Resource Policy in India. A staunch opponent of the TRIPS agreement, Shiva points out that under the treaty "intellectual property rights are recognized only as private rights" and thus "exclude all kinds of knowledge, ideas, and innovations that take place in the 'intellectual commons'—in villages among farmers [and] in forests among tribespeople . . ."[32] Monsanto and the other companies that helped frame the TRIPS agreement are not unaware of the value of indigenous knowledge. Their gene prospectors are constantly on the prowl in remote villages, questioning village elders, in hopes of locating a special medicinal herb or plant crop with unique genetic properties.

Suman Sahai, director of the Gene Campaign, a non-governmental organization in New Delhi, makes an obvious point, when he says, "God didn't give us 'rice' or 'wheat' or 'potato.' " These were once wild plants that were domesticated over eons of time and patiently bred by generations of farmers. Sahai asks, "Who did all of that work?"[33]

Unfortunately, patent laws only reward individual innovative efforts in scientific laboratories. Collective efforts, passed down from one generation to another, are rationalized as "prior art" and dismissed altogether. It appears to many in the third world that the biotech companies are, in effect, getting a free ride on the back of thousands of years of indigenous knowledge. Corporations browse the centers of genetic diversity, helping themselves to a rich largess of genetic treasures, only to sell back the same in a slightly engineered and patented form, and at a hefty price—all for products that have been freely shared and traded among farmers and villagers for all of human history.

The financial rewards of successful bioprospecting are often quite significant. In 1993, Lucky Biotech Corporation, a Korean pharmaceutical firm, and the University of California were awarded U.S. and international patents for a genetically engineered sweet protein derived from a plant found in West Africa called thaumatin. The thaumatin plant protein is one hundred thousand times sweeter than sugar, making it the sweetest substance on Earth. It has been used by local villagers for centuries as a sweetener for food. The thaumatin protein can be engineered into the genetic code of fruits and vegetables, providing a low-calorie sweetener. With the market for low-calorie sweeteners nearing a billion dollars in the U.S. market alone, thaumatin is likely to become a cash cow in the coming years. Meanwhile, villagers in West Africa will not share in the good fortune, even though their ancestors are the real discoverers of thaumatin.[34]

In an effort to ease the growing tensions between global companies and Southern countries, a number of international institutions and private companies have proposed plans to share a portion of their commercial gain from new patents on biotech products with the host countries, local peoples, and other interested parties. The International Plant Genetic Resources Institute (IPGRI), headquartered in Rome, has proposed that companies seeking to market agricultural products derived from germplasm stored in international agricultural research centers be required to negotiate royalty arrangements with the source countries. IPGRI's Director General Geoffrey Hawtin was less than enthusiastic, however, in announcing the plan, complaining that the proposed royalty arrangements would undermine research efforts to increase world food production by restricting the free flow of germplasm. More fervent advocates of the proposed plan took a different spin—arguing that providing financial incentives to source countries would encourage conservation efforts and help preserve genetic diversity.[35]

In the United States, the National Cancer Institute (NCI), which annually collects more than six thousand plant and marine organisms from around the world, has drafted an agreement which says that it "recognizes the need to compensate source country organizations and peoples in the event of commercialization of a drug derived from the organism collected."[36] The NCI has signed a dozen such agreements with countries, including the Philippines, Belize, Indonesia, and Madagascar. Upon closer inspection, however, NCI's agreement appears far less substantial than it would first appear. The Institute says only that if an organism collected by NCI is licensed out to a pharmaceutical company, the company will be required to enter into agreements that "address the concern . . . that pertinent agencies, institutions and/or persons receive royalties and other forms of compensation, as appropriate."[37]

Private companies have also entered into agreements with source countries, local institutions, and indigenous peoples, to "share" a portion of the gains from patented products. The most controversial and highly publicized venture was initiated by Merck & Co. in Costa Rica. The giant pharmaceutical company entered into an agreement with a local research organization, the National Biodiversity Institute, in which Merck agreed to pay a little more than one million dollars to the organization in return for securing the company's potentially valuable plant, microorganism, and insect samples. Critics liken the deal to European settlers giving American Indians trinkets worth a few dollars in return for exclusive ownership of the island of Manhattan. A company boasting $4 billion in revenue, purchasing bioprospecting rights in a country with one of the richest reservoirs of plant and animal life on Earth, for a trifling one million dollars, is little more than tokenism.[38] The recipient organization, on the other hand, is granting a right to bio-prospect on land that it has no historic claim to in the first place. Meanwhile the indigenous peoples, who might have legitimate claim to negotiate a transfer of germplasm, are locked out of the Merck agreement, casting further doubt on the integrity of Merck's arrangement.

To summarize, transnational companies argue that patent protection is essential if they are to risk financial resources and years of research and development bringing new and useful products to market. Southern countries, however, argue that the real research and development effort takes place years before by villagers and peasant farmers who isolate, enhance, and preserve valuable herbs and plant crops. That being the case, they claim some form of compensation for their contribution to the biotech revolution. Despite their differences, both parties share a fundamental assumption—a willingness to commercially enclose the global gene pool for the first time and transform it into a commodity that can be priced in the marketplace.

A growing number of non-governmental organizations, as well as some countries, are beginning to take a third position, arguing that the gene pool ought not to be for sale, at any price—that it should remain an open commons and continue to be used freely by present and future generations. They cite precedent in the recent historic decision by the nations of the world to maintain the continent of Antarctica as a global commons free of commercial exploitation. It would be impossible, they argue, to affix a true market value on indigenous knowledge. How would one even begin to estimate the value of thousands of years of collective knowledge or, for that matter, determine the future worth of a unique genetic trait? Equally important, how would one ever determine who is to be the legitimate recipient of royalty payments, when the knowledge gained is a shared knowledge, more often than not the result of multiple contributions between tribes and people over the millennia?

Some even argue that, far from being a progressive force, patent protection has the opposite effect—stifling the free exchange of information so essential to improving the human condition. Dr. Martin Kenney, professor of human and community development at the University of California at Davis, says that in many instances,

> the fear of being scooped or seeing one's work transformed into a commodity can silence those who presumably are colleagues. To see a thing that one produced turned into a product for sale by someone over whom one has no control can leave a person feeling violated. The labor of love is turned into a plain commodity—the work now is an item to be exchanged on the basis of its market price. Money becomes the arbiter of a scientific development's future.[39]

A Harvard study on the linkages between university researchers and biotech firms reached similar conclusions. A team headed by Dr. David Blumenthal of Harvard's Center for Health Policy and Management surveyed 550 companies engaged in biotech research and found that more than 20 percent of their research and development funding went to university researchers. The close relationship between university scientists and the biotech industry, according to the study, has had a chilling effect on the sharing of research. Forty-one percent of the biotech firms reported at least one trade secret arising out of their university-funded work. Trade secrets were defined as proprietary information "protected through systematic attempts to prevent disclosure, including prohibiting publication of research results."[40] The study also found that "biotechnology faculty with industry support were four times as likely as other biotechnology faculty to report that trade secrets had resulted from their university research."[41] Many of the faculty engaged in corporate-sponsored biotech research are also principals in the same companies—serving on boards of directors, or being paid as consultants. Most of the top-flight researchers enjoy significant equity holdings in the companies.[42]

A similar study conducted by Dr. Sheldon Krimsky, professor of urban and environmental policy at Tufts University, in the late 1980s found that an incredible 37 percent of the biotechnology scientists who were members of the prestigious National Academy of Sciences—a body that advises Congress and the federal government on important matters relating to science policy—had "industry affiliations," casting serious doubt on their objectivity in questions relating to biotech science policies.[43] Critics like Krimsky argue that the commercially driven nature of biotech research has seriously undermined the traditional collegial laboratory atmosphere of sharing ideas and has slowed cooperative efforts to find solutions to problems. Secrecy has become paramount in a commercially directed system where the reward for research is no longer simply the respect and admiration of peers and contribution to knowledge but rather the patenting of potentially lucrative inventions. The intimate collaboration between industry and the research community, says biochemist Keith Yamamoto of the University of California at San Francisco, "has affected adversely both communication and morale within the academic setting."[44]

## Human Beings as Intellectual Property

The debate over life patents has taken on even greater urgency of late with increasing reports of scientific institutions as well as pharmaceutical and biotech companies bioprospecting the human genome itself in remote regions of the world. The Human Genome Diversity Project, a scientific effort headed by Dr. Luigi Luca Cavalli-Sforza, a population geneticist and professor emeritus at Stanford University, has been the subject of growing controversy after it became public that the group planned to take blood samples from the world's five thousand linguistically distinct populations in order to assess their genetic makeup and search for any unique genetic traits they may have that might prove important and useful in the future.

Project sponsors hope that by sampling the genotypes of the few groups of indigenous peoples that have remained isolated from the rest of the outside world, they will find some "genetic surprises" that could be a boon to humankind in its search for new ways to improve the genetic makeup of the human race. Dubbed the "vampire project" by its critics, Cavalli-Sforza defends the research, saying it is important to seek out whatever remaining genetic variety exists before it's irretrievably lost, either as a result of the extinction of these populations, or through their melding into the general human population. Cavalli-Sforza says that while his personal view is that there should be no patents on DNA, he feels that the potential commercial value of the genetic information emerging from the Human Genome Diversity Project may make such notions impractical. He suggests, therefore, that "in the unlikely event that there is some gene that becomes commercially valuable, the people who donated it—not the individual, but the group—should somehow share in the advantages."[45]

The concerns of critics came alive in 1993, when the Rural Advancement Foundation International, a non-governmental organization, discovered that the U.S. government had sought both U.S. and international patents on a virus derived from the cell line of a twenty-six-year-old Guaymi Indian woman from Panama. A NIH researcher had taken a blood sample from the woman and developed the cell line. The Guaymi cell line was of particular interest to NIH researchers because members of this remote Indian community carry a unique virus that stimulates the production of antibodies that scientists believe might be useful in AIDS and leukemia research.[46]

Upon learning about the patent application, representatives of the Guaymi General Congress in Panama waged a public protest. Isidro Acosta, president of the Guaymi General Congress, said at the time that he was shocked that an august scientific body like the NIH could so wantonly violate the genetic privacy of his tribe and that the U.S. government, without advising the Guaymi of its intentions, could then seek to patent a genetic trait of the Guaymi and profit from their biological inheritance in the global marketplace.

> I never imagined people would patent plants and animals. It's fundamentally immoral, contrary to the Guaymi view of nature, and our place in it. To patent human material . . . to take human DNA and patent its products . . . that violates the integrity of life itself, and our deepest sense of morality.[47]

The public protest forced the U.S. government to withdraw its patent application. The controversy over patenting genes from indigenous peoples flared again, however, several months later when the U.S. government filed two additional patent claims in the United States and Europe for cell lines taken from citizens of the Solomon Islands and Papua New Guinea. When the government of the Solomon Islands issued a protest, then Secretary of Commerce, the late Ron Brown, responded curtly,

> Under our laws, as well as those of many other countries, subject matter relating to human cells is patentable and there is no provision for considerations relating to the sources of the cells that may be the subject of a patent application.[48]

In March of 1995, the U.S. Patent Office issued a patent for the Papua New Guinea Human T-lymphotrophic virus (HTLV-1) to the U.S. Department of Health and Human Services, making it the first human cell line from an indigenous population to be patented.[49] Angry over the United States action, a group of South Pacific Island nations prepared a joint proposal and communique that would make their sovereign space a "patent-free zone."[50] The U.S. government quietly dropped the patent claim in 1996.

The Indian government has also expressed its deep misgivings about research efforts to secure blood samples from its many ethnic populations. India, with its diverse cultures and inbred populations, is considered to be an ideal setting for gene prospecting. "Name any genetic disorder and we have the mutations," says Samir Brahmachari, professor of molecular biophysics at the Indian Institute of Science in Bangalore.[51] For example, the Onge tribes of India have a small Y chromosome and low sperm count. In southern India, a small community of seven hundred families suffer from osteoarthritis and dwarfism. In west Bengal, where cholera is an ever-present threat, a large group of people appear to be immune to the disease. Scientists are now searching for the gene or genes that might confer the genetic advantage in hopes of finding a new form of treatment for cholera.[52] K. Suresh Singh, until recently chief of the Anthropological Survey of India, reports being swamped by proposals from researchers in the United States and other countries who would like to conduct field studies with various tribes and ethnic groups throughout India.[53]

In January of 1996, the Indian Society of Human Genetics (ISHG) issued a set of guidelines calling for a ban on the transport of "whole blood, cell-lines, DNA, skeletal and fossil material" pending formal agreements between collaborating parties. The ISHG made it clear that any such agreement should specify "the objectives of the project and the anticipated scientific material and economic benefits and the manner that they are to be shared both at present and in the future."[54] The action came in the wake of revelations that the NIH was illegally securing DNA and blood samples from patients at private eye hospitals in India without obtaining the appropriate authorization required to take samples out of the country. U.S. researchers from the NIH's National Eye Institute were searching for genes that cause retinitis pigmentosa, or night blindness. The Indian Council of Medical Research (ICMR) said it should have been notified of the research in advance, as Indian law forbids the export of biological material without permission from the ISHG.[55]

The rush to locate and patent commercially valuable genes has become so intense that researchers are being courted with hard cash to assist in the gene prospecting. Kiren Kucheria, a geneticist at the All India Institute of Medical Sciences in New Delhi, reports being offered $20,000 for blood samples from two of her patients suspected of harboring a special gene that plays a role in a disorder called mullerian ductagenesis, which is characterized by the absence of a uterus.[56] She refused to cooperate.

Securing the rights to a particular cell line or gene can translate into potentially large financial rewards down the line. Recently, the biotech firm Amgen paid Rockefeller University $20 million for the right to develop products from a gene that may play a role in determining obesity.[57]

It seems there is no place on Earth too remote for the gene-hunters to go. In April of 1997, *The Los Angeles Times* reported on a scientific expedition led by Dr. Noe Zamel, a University of Toronto Medical geneticist, and financed by Sequana Therapeutics of La Jolla California. Sequana is one of a handful of new biotech start-up companies dedicated to gene prospecting. These "genomics" firms are on the commercial frontier of the emerging biotech revolution.

The Sequana team traveled, by way of a South African navy supply ship, to the tiny volcanic island of Tristan da Cunha, a little strip of forty square miles in the middle of the Atlantic Ocean. Often referred to as the world's "loneliest island," its few hundred residents descended from British sailors who first arrived there in 1817. What makes the small inbred local population interesting to Zamel and his team is that half of them suffer from asthma. Scientists hope to find the gene or genes responsible and patent them.

Company scientists took blood samples from 270 of the island's three hundred residents and later reported that it had located two "candidate genes" responsible for asthma. However, thus far they have refused to share their findings with other researchers in the field, giving rise to charges that the company is putting commercial considerations above collaborative efforts to find a cure for the disease. For their part, genomic firms like Sequana acknowledge that they're in business to commercially exploit the human genome and could not hope to turn a profit if they couldn't keep their research proprietary—at least until it was patented.[58] Sequana and other genomic firms maintain that market incentives are the best and most efficient way to advance the research. Others aren't so sure. "The issue is a nightmare," said one geneticist working for a large pharmaceutical company, who spoke off the record to *Nature* magazine's editor.

Foreign nationals aren't the only ones whose cell lines and genomes are being patented by commercial firms in the United States. In a precedent-setting case in California, an Alaskan businessman named John Moore found his own body parts had been patented, without his knowledge, by the University of California at Los Angeles (UCLA) and licensed to the Sandoz Pharmaceutical Corporation. Moore had been diagnosed with a rare cancer and underwent treatment at UCLA. An attending physician and university researcher discovered that Moore's spleen tissue produced a blood protein that facilitates the growth of white blood cells that are valuable anti-cancer agents. The university created a cell line from Moore's spleen tissue and obtained a patent on their "invention" in 1984. The cell line is estimated to be worth more than $3 billion. Moore subsequently sued the University of California claiming a property right over his own tissue. In 1990 the California Supreme Court ruled against Moore, holding that Moore had no property right over his own body tissues. Human body parts, the Court argued, could not be bartered as a commodity in the marketplace. However, the Court did say that the "inventors" had a responsibility to inform Moore of the commercial potential of his tissue and, for that reason alone, had breached their fiduciary responsibility and might be liable for some kind of monetary damages. Still, the Court upheld the primary claim of the university that the cell line itself, while not the property of Moore, could justifiably be claimed as the property of UCLA. The irony of the decision was captured by Judge Broussard, in his dissenting opinion. He wrote,

> . . . the majority's rejection of plaintiff's conversion cause of action does *not* mean that body parts may not be bought or sold for research or commercial purpose or that no private individual or entity may benefit economically from the fortuitous value of plaintiff's diseased cells. Far from elevating these biological materials above the marketplace, the majority's holding simply bars *plaintiff,* the source of the cells, from obtaining the benefit of the cell's value, but permits *defendants,* who allegedly obtained the cells from plaintiff by improper means, to retain and exploit the full economic value of their ill-gotten gains free of . . . liability.[59]

The extraordinary implications of privatizing the human body—parceling it out in the form of intellectual property to commercial institutions—is illustrated quite poignantly in the case of a patent awarded by the European Patent Office to a U.S. company named Biocyte. The patent gives the firm ownership of all human blood cells which have come from the umbilical cord of a newborn child and are being used for any therapeutic purposes. The patent is so broad that it allows this one company to refuse the use of any blood cells from the umbilical cord to any individual or institution unwilling to pay the patent fee. Blood cells from the umbilical cord are particularly important for marrow transplants, making it a very valuable commercial asset. It should be emphasized that this patent was awarded simply because Biocyte was able to isolate the blood cells and deep-freeze them. The company made no change in the blood itself. Still, the firm now possesses commercial control over this part of the human body.[60]

A similarly broad patent was awarded to an American firm, Systemix Inc. of Palo Alto, California, by the U.S. PTO, covering all human bone marrow stem cells. This extraordinary patent on a human body part was awarded despite the fact that Systemix had done nothing whatsoever to alter or engineer the cells. Some, even in the medical establishment, were stunned by the PTO decision. Dr. Peter Quesenberry, the medical affairs vice chairman of the Leukemia Society of America, quipped, "Where do you draw the line? Can you patent a hand?"[61]

On the other hand, many in the molecular biology field see no apparent ethical problem or moral dilemma in gene prospecting and patenting and are laying claims to vast regions of the human genome. Arnold Slutsky of the Samuel Lunenfeld Institute of the Mount Sinai Hospital of Toronto asks rhetorically, "If a journalist writes an article on a family, and then wins a Pulitzer Prize, does he give the family a percentage of his winnings?"[62]

The entrepreneurial scramble to patent the genome of the human family has picked up substantial momentum over the last several years, in large part because of the quickened pace of mapping and sequencing the approximately 100,000 genes that make up the human genome. As soon as a gene is tagged, its "discoverer" is likely to apply for a patent, often before even knowing the function or role of the gene. In 1991, J. Craig Venter, then head of the NIH Genome Mapping Research Team, resigned his government post to head up a genomics company funded by a venture capital fund for more than $70 million.[63] At the same time, Venter and his colleagues filed applications seeking patents on more than two thousand human brain genes. Many researchers working on the Human Genome Project were shocked and angry, charging Venter with attempting to profit off research initially paid for by American taxpayers. A number of scientists were upset because Venter sought patents on genes before even knowing their function. Nobel Laureate Dr. James Watson, the co-discoverer of the DNA double helix and the former head of the Human Genome Project, called the Venter patent claims "sheer lunacy."[64] Still, Venter and the many other scientists and genomic firms that followed have zealously pursued their goal of laying claim to as much of the human genome as possible. It's likely that within less than ten years, all one hundred thousand or so genes that comprise the genetic legacy of our species will be patented, making them the exclusive intellectual property of global pharmaceutical, chemical, agribusiness, and biotech companies.

## The Backlash

The idea of private companies laying claim to thousands of human genes as their exclusive intellectual property has resulted in growing protests around the world, especially by groups immediately and intimately affected by the patent claims. In May of 1994, a coalition of hundreds of women's organizations from more than forty nations publicly announced their collective opposition to the attempt by Myriad Genetics, a U.S. biotechnology firm, to patent the discovery of a gene that causes breast cancer in some women who have a history of breast cancer in their families. The coalition was assembled by The Foundation on Economic Trends. The women made it clear that while they were not opposed to the screening test Myriad had developed to detect the existence of the BRCA1 gene in at-risk women—although some had serious reservations as to how broadly administered the test should be—they were opposed to the company laying claim to the gene itself. The women argued that the breast cancer gene was a product of nature and not a human invention, and therefore, should not be patentable. If Myriad were to be allowed exclusive rights over the commercial use of the breast cancer gene, the women claimed, it would result in more expensive screening tests, barring many poor women from screening. In addition, exclusive control of the gene by Myriad Inc. might impede further research on breast cancer, as the cost of gaining access to the gene might be too expensive, especially for academic researchers. The biotech industry shot back, claiming that patent protection was an essential incentive for companies to invest valuable time and financial resources in research and development efforts.[65]

The issue of life patents has been the subject of growing scrutiny in the legislative arena. In the United States Congress several bills designed to limit the reach of life patents or impose a moratorium on the issuing of any further patents, pending study and debate, have passed both the House and Senate. Yet, to date, no agreed-upon legislation has simultaneously cleared both chambers and been enacted into law.

In Europe, the debate over life patents has reached a near fever pitch, at times putting the European Parliament—the legislative body of the European Union (EU)—at loggerheads with both the biotech industry and the European Commission, the administrative oversight body representing the separate nations of the EU. In 1995, the European Parliament rejected a proposed "Life-Patents Directive" to harmonize the various patent regimes of the member countries by aligning them with the broad patent policies on life forms existing in the United States. The Parliament argued against the patenting of human genes, cells, tissue, organs, and embryos on moral, religious, and philosophical grounds, saying that the human genome ought not be reduced to commercial property to be bartered and sold in the open marketplace. The legislative body also emphasized that human genetic material is a fact of nature and therefore must be considered a "discovery" and not an "invention." Finally, the Parliament was adamant in its contention that the granting of monopoly patents would discourage the open exchange of vital information and slow down or forestall cooperative efforts to find cures for diseases and develop new medical treatments.[66]

In the wake of the unexpected defeat of the patent directive, the European Commission joined hands with the biotech industry and submitted a revised directive in 1997 designed to mollify the members of the European Parliament. Supporters of the initiative bolstered their legal and commercial claims with emotional appeals, including the mobilization of disabled teenagers to tell members of Parliament that biotech companies would not invest in life-saving medical treatments and cures if their research was not rewarded with the issuance of patents. In July of 1997, after waging an intensive and unprecedented lobbying campaign, the biotech industry succeeded in convincing the European Parliament to pass a draft of the newly revised patent directive by a vote of 388 to 110.[67]

The debate over life patents is one of the most important issues ever to face the human family. Life patents strike at the core of our beliefs about the very nature of life and whether it is to be conceived of as having intrinsic or mere utility value. A great debate of this kind occurred more than half a millennium ago when the Catholic Church and an emerging merchant and banking class squared off over the question of usury. The Church argued for a "just price" between sellers and buyers and claimed that merchants and bankers could not profit off time by charging rapacious rates of interest because time was not theirs to negotiate. God freely dispenses time and it is only His to give and take. In charging interest, said Thomas Chobham, the medieval scholar, "the usurer sells nothing to the borrower that belongs to him. He sells only time which belongs to God. He can therefore not make a profit from selling someone else's property."[68] The merchants disagreed, and argued that time is money and that charging interest on time was the only way to secure their investments in the market. The Church lost the battle over usury and the charging of interest, and the loss set a hurried course toward market capitalism and the modern age.

In the nineteenth century, another great debate ensued over the issue of human slavery, with abolitionists arguing that every human being has intrinsic value and "God-given rights" and cannot be made the personal commercial property of another human being. The abolitionists' argument ultimately prevailed, and legally sanctioned human slavery was abolished in every country in the world where it was still being practiced. Unfortunately, human slavery, in the form of involuntary indenture, is still a cultural practice in some areas of the world. The inherent rights of women, minorities, children, and animals have become a matter of increasing public concern over the course of the present century as advocates struggle to ensure the triumph of intrinsic values over utility values.

Now, still another grand battle is unfolding on the eve of the Biotech Century, this time over the question of patenting life—a struggle whose outcome is likely to be as important to the next era in history as the debate over usury, slavery, and involuntary indenture have been to the era just passing. In May of 1995, a coalition of more than two hundred religious leaders, including the titular heads of virtually every major Protestant denomination, more than one hundred Catholic bishops, and Jewish, Muslim, Buddhist, and Hindu leaders, announced their opposition to the granting of patents on animal and human genes, organs, tissues, and organisms. The effort was organized by The Foundation on Economic Trends.

The coalition, the largest assemblage of U.S. religious leaders to come together on an issue of mutual interest in the twentieth century, said that the patenting of life marked the most serious challenge to the notion of God's Creation in history. How can life be defined as an invention to be profited from by scientists and corporations when it is freely given as a gift of God, ask the theologians? Either life is God's Creation or a human invention, but it can't be both. Speaking for the coalition, Jaydee Hanson, an executive with the United Methodist Church, said, "We believe that humans and animals are creations of God, not humans, and as such should not be patented as inventions."[69] While not all the religious leaders oppose "process" patents for the techniques used to create transgenic life forms, they are unanimous in their opposition to the patenting of the life forms and the parts themselves. They are keenly aware of the profound consequences of shifting authorship from God to scientists and transnational companies and are determined to hold the line against any attempt by "man" to stake his own claim as the prime mover and sovereign architect of life on Earth.

# Three A Second Genesis

While the biotech revolution will reshape the global economy and remake our society, it is likely to have an equally significant impact on the Earth's environment. The new technologies of the Genetic Age allow scientists, corporations, and governments to manipulate the natural world at the most fundamental level—the genetic components that help orchestrate the developmental processes in all forms of life. In this regard, it is probably not overstating the case to suggest that the growing arsenal of biotechnologies is providing us with powerful new tools to engage in what will surely be the most radical experiment on the Earth's life forms and ecosystems in history. Imagine the wholesale transfer of genes between totally unrelated species and across all biological boundaries—plant, animal and human—creating thousands of novel life forms in a brief moment of evolutionary time. Then, with clonal propagation, mass-producing countless replicas of these new creations, releasing them into the biosphere to propagate, mutate, proliferate, and migrate, colonizing the land, water, and air. This is, in fact, the great scientific and commercial experiment underway as we turn the corner into the Biotech Century.

Reseeding the biosphere with a laboratory-conceived second Genesis is the third strand in the operational matrix of the Biotech Century. Transnational companies are making a transition out of chemicals and into the life sciences, positioning themselves to control global markets in the Age of Biology. Typical of the new trend is the bold decision by the Monsanto Corporation, long a world leader in chemical products, to sell off its entire chemical division in 1997 and anchor its research, development, and marketing in biotech-based technologies and products. (Monsanto retained control of its agrochemical division, however.) Global conglomerates are rapidly buying up biotech start-up companies, seed companies, agribusiness and agrochemical concerns, pharmaceutical, medical and health businesses, and food and beverage companies, creating giant life-sciences complexes from which to fashion a bioindustrial world.

Several factors have combined to create what industry analysts are calling a global "life industry." The relaxing of trade restrictions with new global trade agreements, including the General Agreement on Tariffs and Trade (GATT), Maastricht, and the North American Free Trade Agreement (NAFTA), the new ease of managing and integrating far-flung business interests by way of computers and advanced telecommunications technology, and the spectacular advances in biotechnologies have all helped spur the creation of a new kind of global commerce that trades in "life products" of every kind. The consolidation of the life sciences industry by global commercial enterprises rivals the consolidations, mergers, and acquisitions going on in the other great technology arena of the twenty-first century, computers, telecommunications, entertainment, and information services, although much less attention has been focused on the life sciences companies in the media and public policy.

The concentration of power is impressive. The top ten agrochemical companies control 81 percent of the $29 billion global agrochemical market. Ten life science companies control 37 percent of the $15 billion per year global seed market. The world's ten major pharmaceutical companies control 47 percent of the $197 billion pharmaceutical market. Ten global firms now control 43 percent of the $15 billion veterinary pharmaceutical trade. Topping the life science list are ten transnational food and beverage companies whose combined sales exceeded $211 billion in 1995.[1]

The life sciences companies are anxious to exploit the enormous potential of the new biotechnologies and are devoting considerable funds to research and development and licensing agreements. An estimated 7.5 billion dollars annually is currently invested in-house on biotechnology programs.[2] Global pharmaceutical companies also spent more than 3.5 billion dollars in 1995 buying up biotech firms. In addition to outright acquisitions, these same life science conglomerates spent approximately 1.6 billion dollars in 1995 in licensing arrangements with biotech firms.[3]

Some of the largest life sciences companies are strategically positioned to control much of the global bioindustrial market in the coming century. Novartis, a giant new global firm resulting from the $27 billion merger of two Swiss companies, the pharmaceutical company Sandoz and the agrochemical company Ciba-Geigy, is typical of the trend toward corporate consolidation in the life industry in the new era of global commercial markets. Novartis is the world's largest agrochemical company, the second largest seed company, the second largest pharmaceutical company, and the fourth largest veterinary medicine company.[4] The company is also staking out claims in the new field of human genetic medicine. In 1995, Sandoz—now Novartis—purchased Genetic Therapy Inc. for $295 million. The Maryland-based firm held a license on a broad patent covering the technique for removing cells from a patient, modifying their genetic makeup, and returning them to the patient's body.[5] The acquisition guaranteed Novartis a foothold in the fledgling science of human gene therapy.

Other global life science companies are close on the heels of Novartis, busily buying up related life science ventures around the world. The Monsanto Corporation acquired Holden's Foundation Seeds in 1997 for $1.2 billion. More than 35 percent of the corn acreage planted in the United States is derived from germplasm developed by Holden. Monsanto also holds a 40 percent share in a second major seed company, DeKalb. Other recent purchases include Asgrow, a leading soybean company, and Agracetus and Calgene, two high-profile agricultural biotech firms. Dow Elanco has purchased a 65 percent share of Microgen, a biotech company that holds a number of potentially valuable patents in the agricultural field. Du Pont, the world's fifth largest agrochemical company, purchased a 20 percent share in Pioneer Hi-Bred, the world's largest seed company, in 1997 for $1.7 billion. Du Pont also acquired Protein Technologies International from Ralston Purina for $1.5 billion. AgrEvo purchased Plant Genetic Systems in 1996 for $725 million.[6]

Meanwhile, the pharmaceutical giants are purchasing equity holdings and establishing research agreements with many of the human genomic companies, convinced that the future of pharmaceuticals and medicine rests with the data being collected on human genes and genetic traits, dispositions, and disorders. Schering Plough acquired Canji for $54.4 million in 1996. Bayer, Novartis, and Eli Lilly have developed commercial relationships with Myriad Inc., the U.S.-based genomic company that, among other things, discovered a breast cancer gene. The company is expected to begin commercial screening for the gene within the next year. Pfizer, Pharmacia, and Upjohn have invested in Incyte, a U.S. firm whose database is supposed to contain the partial sequences for nearly 100,000 genes. Eli Lilly has entered into an agreement with Millennium Pharmaceuticals, a company doing genomic research on atherosclerosis. Corange International made a $100 million research agreement with Gene Medicine, a U.S. genomic firm. Glaxo Wellcome has a five-year agreement with Sequana Therapeutics to advance work on a gene associated with obesity and diabetes. Synthelabo of France has bought a $9.7 million equity investment in Genset of France as well as entering into a $69 million research agreement with the genomic firm. SmithKline Beecham has made a $125 million research agreement with Human Genome Sciences Inc., a U.S.-based genomic firm.[7]

Corporate leaders in the new life sciences industry promise an embarrassment of riches and much more—they offer a door to a new era of history where evolution itself becomes subject to human authorship. Critics worry that the reseeding of the Earth with a second Genesis could lead to a far different future—a biological Tower of Babel spreading chaos throughout the biological world and, in the process, drowning out the ancient language of evolution.

Genetic pollution is already appearing and is likely to spread in the coming Biotech Century, destroying habitats, destabilizing ecosystems, and diminishing the remaining reservoirs of biological diversity on the planet. This newest form of pollution is also likely to create serious and potentially catastrophic health risks for many of the Earth's animal species and human beings.

Molecular biologists and industry spokespersons argue that the Biotech Century will unfold without serious environmental consequences to the planet. Their assurances, however, are being met with increasing skepticism. There is not a single instance in history in which the introduction of a major technological innovation has had only benign consequences for the natural world. New technologies allow human beings to exploit and expropriate nature for short-term gain, but always at the expense of polluting, depleting, and destabilizing some portion of the biosphere in the process. The power to transform, remake, and exploit nature in wholly new ways virtually guarantees that the biotech revolution will inflict its own form of damage on the Earth's environment. Indeed, genetic pollution is likely to pose at least as significant a threat to the biosphere in the coming century as petrochemicals have in the current century.

Human beings have been remaking the Earth for as long as we have had a history. Up to now, however, our ability to create our own second Genesis has been tempered by the restraints imposed by species boundaries. We have been forced to work narrowly, continually crossing close relatives in the plant or animal kingdoms to create new varieties, strains, and breeds. Through a long historical process of tinkering and trial and error we have redrawn the biological map, creating new agricultural products, new sources of energy, more durable building materials, life-saving pharmaceuticals, and other useful products. Still, in all this time, nature dictated the terms of engagement.

The new gene splicing technologies allow us to break down the walls of nature, making the very innards of the genome vulnerable to a new kind of human colonization. Transferring genes across all biological barriers and boundaries is a technological tour de force, unprecedented in human history. We are experimenting with nature in ways never before possible, creating unfathomable new opportunities for society and grave new risks for the environment.

The potential dangers first came to light in a dramatic set of genetic engineering experiments conducted by the National Institute of Allergy and Infectious Disease in the late 1980s. Anxious to find an animal model suitable for the study of AIDS, researchers introduced the human AIDS virus genome into mouse embryos by micro-injection. The mice were born, expressing the human immunodeficiency virus in every cell of their bodies. For the first time, scientists had successfully introduced the genetic instructions for a virus that inflicts humans into the genetic code of another animal. Equally important, a number of the mice in subsequent generations carried the HIV virus.

Critics warned of the slim but very real chance of the mice accidentally escaping the laboratory into the open environment where they might mate with feral mice, creating a new and deadly reservoir for AIDS in the animal world. The scientists conducting the research dismissed the warnings as "alarmist." Still, the National Institutes of Health (NIH) was sufficiently worried that it housed the AIDS mice in a stainless-steel glove box surrounded by a moat of bleach that, in turn, was enclosed in a biosafety level-four facility—the highest bio-security facility that exists.

Then in February of 1990, Dr. Robert Gallo, the co-discoverer of the AIDS virus, and a team of scientists published the results of a study they conducted on the suitability and advisability of using the AIDS mouse as an animal research model in the journal *Science.* Gallo and his colleagues reported that the AIDS virus carried by the mice could combine with other mouse viruses, resulting in the creation of a new and more virulent form of the AIDS virus—a "super" AIDS.

According to the report, the new super AIDS virus strain acquired new biological characteristics, "including the ability to reproduce more rapidly than it normally does and to infect new kinds of cells." Even more frightening, the new virus might spread by "novel routes" including transmission through air. The authors of the report concluded by urging a "prompt and critical evaluation of the experimental data obtained from some animal models for HIV-1 and of their suitability for use in the study of new antiviral therapeutic or prophylactic approaches."[8]

## A New Environmental Threat

The story of the super AIDS mouse is a cautionary tale of the power unleashed by the new recombinant DNA technology and the unanticipated consequences that can result from combining genetic material across previously impenetrable species boundaries. Virtually every genetically engineered organism released into the environment poses a potential threat to the ecosystem. To appreciate why this is so, we need to understand why the pollution generated by genetically modified organisms is so different from the pollution resulting from the release of petrochemical products into the environment.

Genetically engineered organisms differ from petrochemical products in several important ways. Because they are alive, genetically engineered organisms are inherently more unpredictable than petrochemicals in the way they interact with other living things in the environment. Consequently, it is much more difficult to assess all of the potential impacts that a genetically engineered organism might have on the Earth's ecosystems.

Genetically engineered products also reproduce. They grow and they migrate. Unlike many petrochemical products, it is difficult to constrain them within a given geographical locale. Once released, it is virtually impossible to recall genetically engineered organisms back to the laboratory, especially those organisms that are microscopic in nature. For all these reasons, genetically engineered organisms may pose far greater long-term potential risks to the environment than petrochemicals.

The risks in releasing novel genetically engineered organisms into the biosphere are similar to those we've encountered in introducing exotic organisms into the North American habitat. Over the past several hundred years, thousands of non-native organisms have been brought to America from other regions of the world. While many of these organisms have adapted to the North American ecosystems without severe dislocations, a small percentage of them have run wild, wreaking havoc on the flora and fauna of the continent. Gypsy moth, Kudzu vine, Dutch elm disease, chestnut blight, starlings, and Mediterranean fruit flies come easily to mind. The mongoose, introduced into Hawaii from India to control rodents who were damaging the sugar cane crop, became an environmental nightmare, devouring a wide range of native animals and, in the process, destabilizing the ecosystems of the islands. The zebra mussel, a native of Europe, migrated to North America by attaching itself to ships and has become a formidable pest in the Great Lakes, blocking up water pipes at filtration plants and edging out native species in the Great Lakes region.[9] Each year the American continent is ravaged by these non-native organisms, with destruction to plant and animal life running into the billions of dollars.

Whenever a genetically engineered organism is released, there is always a small chance that it too will run amok because, like non-indigenous species, it has been artificially introduced into a complex environment that has developed a web of highly integrated relationships over long periods of evolutionary history. Each new synthetic introduction is tantamount to playing ecological roulette. That is, while there is only a small chance of it triggering an environmental explosion, if it does, the consequences could be significant and irreversible.

Global life-sciences companies are expected to introduce thousands of new genetically engineered organisms into the environment in the coming century, just as industrial companies introduced thousands of petrochemical products into the environment over the course of the past two centuries. While many of these genetically engineered organisms will be benign, sheer statistical probability suggests that at least a small percentage will prove to be dangerous and highly destructive to the environment.

For example, scientists are considering the possibility of producing a genetically engineered enzyme that could destroy lignin, an organic substance that makes wood rigid. They believe there might be great commercial advantage in using this genetically modified enzyme to clean up the effluent from paper mills or for decomposing biological material for energy. But if the bacteria containing the enzyme were to migrate offsite, it could well end up destroying millions of acres of forests by eating away at the substance that provides trees with their rigidity.

A number of experiments are already under way to release genetically engineered animals into the environment, including predator insects that will prey on noxious insects and genetically engineered fish with growth hormone and "antifreeze" genes inserted into their genetic codes to allow them to grow faster and bigger and be able to tolerate colder waters. Dr. Bernard Rollin, a professor of physiology, biophysics, and philosophy at Colorado State University, asks,

> If we can take animals whose characteristics are well-known, well-understood, and reasonably predictable and put them into environments that are familiar, and we still occasion disaster—sometimes disaster that we can't reverse—how much more likely are we to do so with new organisms, whose traits we do not yet understand?[10]

The Ecological Society of America raised the question of risks associated with the release of genetically engineered organisms into the environment in a special report authored by some of the country's distinguished ecologists. The Society acknowledged that while many new genes introduced into animals will likely reduce fitness and weaken survivability in the wild, there will be exceptions.[11]

For example, fish genetically engineered to increase the efficiency of food conversion, tolerate cold and salinity, and be disease resistant might have a selective advantage if they escaped into open waters.[12] Although most genetically engineered fish are being designed to live in commercial tanks and fish farms, ecologists note a long history of accidental escapes into open waters as a result of unanticipated flooding. Once loose in local aquatic ecosystems, the transgenic fish could potentially out-compete native fish species and create havoc. Even the accidental release of sterile male fish could pose unanticipated problems. If the males were larger and stronger as a result of the addition of a gene to produce increased growth hormone, they might secure easier access to female eggs, crowding out the native males, and because they are sterile, seriously deplete the indigenous fish population.[13]

The long-term cumulative impact of thousands of introductions of genetically modified organisms could well exceed the damage that has resulted from the release of petrochemical products into the Earth's ecosystems. With these new biologically based products, the damage is not easily containable, the destructive effects continue to reproduce, and the organisms cannot be recalled, making the process irreversible.

Consider the first government-approved release of a genetically engineered organism into the open environment. In the early 1980s researchers at the University of California modified a bacteria called *Pseudomonas syringae.* This particular bacterium is found in its naturally occurring state in temperate regions all over the world. Its most unique attribute is its ability to nucleate ice crystals. Using recombinant DNA technology, University of California researchers found a way to delete the genetic instructions for making ice from the bacteria. This new genetically modified *P. syringae* microbe is called ice-minus.[14]

Scientists were excited about the long-term commercial possibilities of ice-minus in agriculture. Frost damage has long been a major problem for American farmers. The chief culprit has been *P. syringae,* which attaches itself to the plants, creating ice crystals. The American corporation financing this research hoped that by spraying massive concentrations of ice-minus *P. syringae* on agricultural crops, the naturally occurring *P. syringae* would be edged out, thus preventing frost damage. The benefits of introducing this genetically engineered organism appeared impressive. It's only when one looks at the long-term ecological costs that problems begin to arise.

To begin with, the first question a good environmental scientist would ask is what role does the naturally occurring *P. syringae* play in nature? The scientists who have studied this particular organism say that its ice-making capacity helps shape worldwide precipitation patterns and is a key determinant in establishing climatic conditions on the planet. Dr. Eugene Odum, professor emeritus of the University of Georgia and one of the nation's best known and respected ecologists, shared his misgivings about releasing the genetically engineered ice-minus bacteria in a letter published in the journal *Science.* Citing research done on *Pseudomonas syringae* by scientists at the National Oceanic and Atmospheric Administration Laboratory in Boulder, Colorado, showing that the bacteria likely play a key role in triggering rainfall, Odum wrote,

> It seems that the lipoprotein coats of this and other species of bacteria found on plants and in detritus when shed and wafted up into the clouds form ideal nuclei for ice formation that is absolutely necessary for rain to fall. . . . If *Pseudomonas syringae* does . . . have a beneficial role in enhancing rainfall, then the ecologist's concern about possible secondary or indirect effects of releases of genetically altered organisms is vindicated. . . .[15]

Dr. Steven Lindow, the University of California researcher conducting the ice-minus experiment, also conceded that "these bacteria may be potentially of critical importance in climatology studies."[16] Even Advanced Genetic Sciences, the Oakland-based biotech firm underwriting the research, acknowledged that "naturally occurring INA [ice-nucleating agent] bacteria may be an important factor influencing precipitation processes."[17] While most ecologists were not overly worried about the controlled release of ice-minus in a single test plot, they were concerned about the long-term effect on worldwide precipitation patterns and climate if ice-minus were to be commercially released over millions of acres of land for a sustained period of time, replacing ice-nucleating *P. syringae* in major agricultural regions of the world.

Prior to U.S. government approval of this first field test of a genetically modified organism in the 1980s, scant attention had been paid to the potential risks involved in this radical new intervention into the natural world. A single congressional hearing was held to address the environmental issues posed by the deliberate release of genetically modified organisms into the environment. A subsequent lawsuit filed by The Foundation on Economic Trends in the U.S. District Court for the District of Columbia sought an injunction against the release of the ice-minus bacteria until such time as the U.S. government conducted and published a thorough environmental impact statement, as required by law. The U.S. District Court granted the injunction and the U.S. Court of Appeals subsequently required the federal government to issue an environmental assessment prior to release of the ice-minus bacteria into the environment.

The environmental impact statement eventually was completed and the field experiment took place, despite the fact that there existed little in the way of a risk assessment science to judge the potential impact of releasing ice-minus, or for that matter, any other genetically modified organism into the open environment. The government, the nation's molecular biologists, and the biotech companies continued the charade, contending that a sufficient body of science existed to measure "the risks," and that adequate regulatory safeguards were in place to assure the safe release of genetically modified organisms. For the most part, the media and the scientific press were complicitous, ridiculing or, worse still, ignoring a growing chorus of criticism from environmental scientists warning of potential dangers in the release of genetically modified organisms.

## Developing a Predictive Ecology

Congressional hearings were held throughout the mid- to late 1980s at which representatives of the Environmental Protection Agency, the National Science Foundation, and the National Academy of Sciences, acknowledged, though half-heartedly, that the new technology lacked a "predictive ecology." The agencies all promised to devote research funds to the furtherance of a risk assessment science, although few funds were ever forthcoming and ecologists were forced to conduct their own research, often on shoestring budgets.

Even today, the U.S. Department of Agriculture (USDA) devotes a mere 1 percent of the funds it allocates to biotechnology research to risk assessment—or a total of only $1 to $2 million per year to study the entire range of environmental issues surrounding the release of transgenic microorganisms, plants and animals into the environment.[18]

The fact is, there has been little interest in pursuing a risk assessment science by the molecular biologists involved in the new gene splicing technologies. There has been even less interest in the subject by biotech companies anxious to gain approval for field tests. Every approved experimental release of a genetically engineered organism virtually assures an infusion of investment funds into corporate coffers and, in the process, often makes instant millionaires of many of the CEOs as well as many of the molecular biologists who serve as principals in the scores of biotech startup companies.

Nowhere have the regulatory shortcomings been more apparent than in the establishment of protocols for field tests of genetically modified organisms. The so called "field tests" were designed to judge potential risk before introducing large-scale commercial releases into the environment. As a risk assessment tool, the field test has to be considered a scientific fiasco. Nonetheless, virtually everyone in the scientific community, the science press, and government went along with the idea without serious reservation, and other countries around the world followed, establishing similar field test protocols.

In 1997, researchers Allison Snow and Pedro Morán Palma laid bare some of the many inherent inadequacies in using field tests as risk assessment tools in one of the very few articles ever published on the subject in a major scientific journal. The two scientists point out that field tests are generally conducted so that "escape of pollen, seeds, and vegetative propagules is unlikely." The researchers go on to say that "gene flow," via pollen, is generally minimized by early harvesting, bagging the flowers, and planting border rows to block the transgenic pollen from escaping the test site. Applicants are even required to explain how they are going to dispose of the plants and seeds after completion of the field test experiment. Thus, according to Snow and Palma, "a major risk associated with commercial production—the escape of fitness-related transgenes via pollen, propagules or seeds—is not addressed in small-scale tests."[19]

Second, because the field test sites are so small—often less than one hundred acres—and the tests themselves are generally limited to only one or two growing seasons, potentially undesirable effects are unlikely to be observed. The question of weeds, insects, and microorganisms building up resistance to herbicide-, pest-, and virus-resistant genes, cannot be adequately addressed in such a small field plot and over such a short period of time. The authors note, "Ecological and evolutionary responses to novel transgenic traits are more likely to occur when hundreds of thousands of acres are dominated by transgenic plants year after year."[20] Then, too, large-scale commercial introductions will take place in countless different ecosystems, each with its own unique soil composition, microorganisms, insects and weather patterns, making the field test virtually useless as an indicator of potential environmental impacts.[21]

Proponents of the field test correctly argue that large-scale field tests over hundreds of thousands of acres, in many diverse ecosystems, for an extended number of growing seasons, might produce more accurate results, but if those results were deleterious, the effects might be irreversible and therefore defeat the very logic of assessing risks before going ahead. They contend that field tests, though inadequate, are better than no tests at all. On the other hand, if the field tests are designed in such a way as to reveal little or nothing of the potential risk that might occur in large-scale commercial releases, then the exercise is little more than a regulatory farce, an elaborate fiction giving the appearance of scientific legitimacy without the substance.

For the most part, the scramble for fame and fortune corrupted the entire regulatory process, with government officials, corporate executives, and molecular biologists working side by side to assure the quick and expedient introduction of genetically modified organisms into the environment, always mindful of the need to maintain the U.S. position of eminence in the emerging new field of biotechnology.

While the public was fed a steady barrage of revised, updated, and reworked regulatory guidelines, giving the appearance of rigorous scientific oversight, the insurance industry quietly let it be known that it would not insure the release of genetically engineered organisms into the environment against the possibility of catastrophic environmental damage because the industry lacks a risk assessment science—a predictive ecology—with which to judge the risk of any given introduction. In short, the insurance industry clearly understands the Kafkaesque implications of a government regime claiming to regulate a technology in the absence of clear scientific knowledge of how genetically modified organisms interact, once introduced into the environment.

Increasingly nervous over the insurance question, one of the biotech trade associations attempted, early on, to raise an insurance pool among its member organizations, but gave up when it failed to raise sufficient funds to make the pool operable. Some observers worried, at the time, and continue to worry—albeit privately—over what might happen to the biotech industry if a large-scale commercial release of a genetically altered organism were to result in a catastrophic environmental event—for example, the introduction and spread of a new weed or pest comparable to Kudzu vine or Dutch elm disease or gypsy moth that might inflict costly damage on flora and fauna over extended eco-ranges, and for a sustained period of time.

Stark images of the nuclear industry haunt the fledgling biotech industry. In the former case, the nation's nuclear utilities agonized over the question of potential liability if a nuclear mishap or meltdown of the kind that eventually befell the Chernobyl nuclear power plant in the former Soviet Union were to occur in the United States. Realizing that potential liability claims would likely greatly exceed their ability to pay, the industry successfully lobbied Congress for the passage of legislation—the Price Anderson Act—that would have the American taxpayers cover all claims beyond a specified amount pledged by the nation's nuclear power companies.

The biotech companies, however, have been reluctant to lobby Congress for a similar piece of legislation for fear of awakening public concern over the new technology. As a result, the question of liability for catastrophic environmental losses remains unresolved, despite the fact that large-scale commercial releases of genetically modified organisms are now being approved for the first time.

Corporate assurances aside, one or more significant environmental mishaps are an inevitability in the years ahead. When that happens, every nation is going to be forced to address the issue of liability. Farmers, landowners, consumers, and the public at large are going to demand to know how it could have happened and who is liable for the damage inflicted. When the day arrives—and it's likely to come sooner rather than later—"genetic pollution" will take its place alongside petrochemical and nuclear pollution as a grave threat to the Earth's already beleaguered environment.

There are already sufficient reasons to be concerned. While the science of risk assessment remains in its infancy, a number of published environmental studies in the major scientific journals in recent years have begun to confirm some of the nagging speculations that critics have entertained for more than fifteen years on the potential adverse impacts of releasing genetically modified organisms into the biosphere.

## Playing Ecological Roulette

Nowhere are the alarm bells going off louder than in agricultural biotechnology, where the industry is moving quickly to make genetically engineered food crops and animals a commercial reality by the end of the first decade of the twenty-first century. Chemical and agribusiness companies are introducing a new generation of transgenic crops into agriculture with hopes of making a wholesale shift into the new genetics revolution. The biotech crops contain novel genetic traits from other plants, viruses, bacteria, and animals, and are designed to perform in ways that could never have been achieved by scientists working with classical breeding techniques. Many of the new gene-spliced crops emanating from the scientific laboratories seem more like creations from the world of science fiction. Scientists have inserted "antifreeze" protein genes from flounders into the genetic code of tomatoes to protect the fruit from frost damage. Chicken genes have been inserted into potatoes to increase disease resistance. Firefly genes have been injected into the biological code of corn plants to serve as genetic markers. Chinese hamster genes have been inserted into the genome of tobacco plants to increase sterol production.[22]

Ecologists are unsure of the impacts of bypassing natural species boundaries by introducing genes into crops from wholly unrelated plant and animal species. The fact is, there is no precedent in history for this kind of "shotgun" experimentation. For more than ten thousand years classical breeding techniques have been limited to the transference of genes between closely related plants or animals that can sexually interbreed, limiting the number of possible genetic combinations. Natural evolution appears to be similarly circumscribed. As a result, there's little or no precedent for what might occur in the wake of a global experiment to redefine the fundamental rules of biological development to suit the needs of market-driven forces. Might the introduction of novel genes into the genomes of traditional food crops create new characteristics that are unpredictable and uncontrollable? The long and short of the matter is, we simply don't know. That's what makes this intervention into the world of agriculture so problematic. It is a high-risk venture with few ground rules and benchmarks to guide the journey. We are flying blindly into a new era of agricultural biotechnology with high hopes, few constraints, and little idea of the potential outcomes.

For example, consider the ambitious plans to engineer transgenic plants to serve as pharmaceutical factories for the production of chemicals and drugs. Foraging animals, seed-eating birds, and insects who live in the soil will be exposed to a range of genetically engineered drugs, vaccines, industrial enzymes, and hundreds of other foreign substances for the first time, with untold consequences. The notion of large numbers of animal species consuming plants and plant debris containing a wide assortment of chemicals that they would normally never be exposed to is an unsettling prospect.[23] Still, little concern has been expressed by molecular biologists and the chemical and pharmaceutical companies that are pushing into this biological frontier.

Much of the current effort in agricultural biotechnology is centered on the creation of herbicide-tolerant, pest-resistant, and virus-resistant transgenic plants. More than a third of all the field releases in 1993–1994 in the Organization for Economic Cooperation and Development nations involved herbicide-tolerant plants, while 32 percent of the field trials involved pest-resistant plants and 14 percent of the test releases were virus-resistant plants.[24]

Herbicide-tolerant crops are a favorite of companies like Monsanto and Novartis that are anxious to corner the lucrative worldwide market for their herbicide products. More than six hundred million pounds of poisonous herbicides are dumped on U.S. farmland each year, most sprayed on corn, cotton, and soybean crops.[25] Chemical companies gross more than $4 billion per year in U.S. herbicide sales alone.[26]

To increase their share of the growing global market for herbicides, chemical companies have created transgenic crops that tolerate their own herbicides. The idea is to sell farmers patented seeds that are resistant to each particular brand of herbicide in the hope of increasing the companies' share of both the seed and herbicide markets. Monsanto's new herbicide-resistant patented seeds, for example, are resistant to its best-selling chemical herbicide, Roundup.

The chemical companies hope to convince farmers that the new herbicide-tolerant crops will allow for a more efficient eradication of weeds. Farmers will be able to spray at any time during the growing season, killing weeds without killing their crops. While companies like Monsanto claim that the widespread adoption of herbicide-tolerant crops will result in a more sparing use of herbicides, many critics are incredulous. They warn that with new herbicide-tolerant crops planted in the fields, farmers are likely to use even greater quantities of herbicides to control weeds, as there will be less fear of damaging their crops in the process of spraying. Chemical companies' claims that increased use of herbicide-resistant plants will lead to less, not more, use of their herbicide rings hollow among ecologists who suspect that chemical companies would not spend large amounts of research and development funds to create transgenic plants with the aim of selling smaller quantities of their herbicides.

The likelihood of increased use of herbicides raises the possibility of weeds developing resistance, forcing an even greater use of herbicides to control the more resistant strains. In one recent study, researchers at the Charles Sturt University in New South Wales found that ryegrass, a common weed in Australia, was becoming increasingly resistant to Monsanto's Roundup and can tolerate nearly five times the recommended dosage before it is killed.[27] Aware of the growing problem of weed tolerance, Monsanto has applied to the regulatory authorities in a number of countries, requesting an increase in the residue limit for its Roundup chemical on crops from six milligrams per kilogram dry weight to twenty milligrams. The potential deleterious impacts on soil fertility, water quality, and beneficial insects that result from the increased use of poisonous herbicides, like Monsanto's Roundup, are a disquieting reminder of the escalating environmental bill that is likely to accompany the introduction of herbicide-tolerant crops.[28]

The new pest-resistant transgenic crops pose similar environmental problems. Chemical companies are readying transgenic crops that produce insecticide in every cell of each plant. A growing body of scientific evidence points to the likelihood of creating "super bugs" resistant to the effects of the new pesticide-producing genetic crops. Several crops, including Novartis's pest-resistant "maximizer corn" and Rohm and Haas's pest-resistant tobacco, are already available on the commercial market.

Virtually all of the pest-resistant crops contain a gene from a naturally occurring soil bacterium—*Bacillus thuringiensis.* The bacterium produces a crystal protein, known as Bt prototoxin. When the toxin is consumed by larvae and insects, it is activated by the insects' stomach acid and destroys their digestive tract. The naturally occurring Bt toxin is used as a biopesticide spray by organic farmers around the world. They rely on it as their chief line of defense against an array of insects including the corn borer and bollworm.

Unlike the naturally occurring bacterial prototoxin, the transgenic toxin has been altered so that it becomes active immediately upon production by the plant. As it does not have to be activated by stomach acids, it can harm a wider range of insects and soil organisms. The transgene also remains toxic up to three times longer in the soil, making it far more lethal than its naturally occurring counterpart.[29]

The unique qualities of pest-resistant transgenic plants make them especially troubling to entomologists and organic farmers, who worry that the widespread use of Bt crops will build resistance among affected insect species, rendering Bt useless as a biopesticide. They have good reason to be concerned. Resistance to *Bacillus thuringiensis* biopesticides first showed up more than a decade ago. Since that time, eight major species of destructive insects have developed resistance to Bt toxin in either laboratory situations or in the environment, including the Colorado potato beetle, the diamondback moth, and the tobacco budworm.[30]

Concern that the use of the Bt transgenic toxin might create a new generation of resistant "super bugs" was heightened in 1996 when an unusually hot and dry growing season in the southern region of the United States triggered an unanticipated series of events in the transgenic cotton crop. Monsanto's transgenic Nu Cotn was grown, for the first time, across the southern agricultural tier. Investors were banking on the success of Monsanto's first large-scale commercial planting to help bolster the fortunes of agricultural biotechnology. Their hopes were soon dashed.

It is widely known that when stressed by heat and drought, plants often curtail their protein synthesis. In the case of Monsanto's Nu Cotn, it appears that the plants reduced their production of Bt toxin as well. The cotton bollworm, however, flourishes in hot, dry conditions. The combination of diminished Bt production levels and increased levels of bollworm infestation occasioned by the hot, dry weather spelled near disaster for Monsanto's "wonder crop." The worm infestation damaged nearly half of the two million acres planted with Nu Cotn, sending investors scurrying. Monsanto instructed farmers to spray the affected fields with conventional chemical pesticides, angering farmers who had paid a premium price for the new transgenic seeds believing they would not have to spend the money or take the time to spray their crop as they had in the past.[31]

The fact that Monsanto's Bt toxin gene did not perform as expected has many scientists worried. Even in field tests, the genetically engineered gene had killed only 80 percent of the bollworms. A 20 percent survival rate virtually assures that resistant strains of "super bugs" will eventually triumph. Professor of entomology Fred Gould of North Carolina State University makes the point that "eighty percent mortality is exactly what researchers use when they want to breed resistant insects."[32] Gould and other ecologists believe that herbicide-resistant and pest-resistant transgenic plants will increase the likelihood of creating new resistant strains of "superweeds" and "super bugs" in the years ahead.

Biotech companies are also working to create plants that can ward off common plant viruses. The new generation of virus-resistant transgenic crops pose the equally dangerous possibility of creating new viruses that have never before existed in nature. Viral coat protein genes are inserted into the genome of plants, conferring resistance to viral infection from the virus from which the gene was lifted. Scientists are still not sure how or why these coat protein genes protect plants from infection by their own viruses. Nonetheless, the process appears to work.

Virus-resistant transgenic crops could be a potential boon for farmers around the world as well as a windfall for biotech companies. On the other hand, concerns are surfacing among scientists and in the scientific literature over the possibility that the coat protein genes could recombine with genes in related viruses that find their way naturally into transgenic plants, creating a recombinant virus with novel features. Researchers have already reported on a number of such recombinations. In one instance, a cauliflower mosaic virus (CaMV) recombined with CaMV genes on a plant chromosome of a transgenic turnip. The prospect of creating new viruses is troubling and raises serious doubts as to the safety and efficacy of releasing virus-resistant transgenic crops into the open environment.[33]

There is also growing concern that a number of the transgenic commercial introductions will themselves become weeds. The likelihood of a transgenic plant becoming a weed is thought, by some ecologists, to be roughly equivalent to the probability of a non-indigenous species becoming a successful weed. Both are novel organisms being introduced, for the first time, into an ecosystem. Neither newcomer nor the habitat have any prior experience accommodating the other. In these situations, scientists generally hold to what they call the ten-ten rule to compute the likelihood of a newcomer becoming a successful invader. That is, it is generally believed that 10 percent of newcomers are likely to successfully establish themselves in their new surroundings and, of those survivors, it is thought that 10 percent of them are likely to ever become significant pests.

Transgenic plants might enjoy slightly better odds than traditional non-indigenous introductions for the reason that many of the transgenic genes inserted into their genomes confer distinct advantages. Herbicide tolerance, pest resistance, and viral resistance are among the transgenic traits that are likely to confer competitive advantage, making transgenic crops potentially formidable invaders in various environments. Dr. Margaret Mellon, a molecular biologist, and Dr. Jane Rissler, a plant pathologist, both with the Union of Concerned Scientists in the United States, cite a number of other transgenic traits currently being tested that might transform a successful crop into an even more successful weed. For example, a transgenic crop containing novel genes that precipitate more rapid germination in cool spring temperatures might grow back as a weed early in the subsequent growing season, creating serious problems for other crops scheduled to be grown in the same field. Transgenes that improve speed of maturation and reproductive capacity might confer similar competitive advantages, allowing a plant to invade surrounding fields, meadows, and forests, and take up permanent residence. A transgenic crop engineered to tolerate colder temperatures could migrate north and successfully invade and colonize new habitats, crowding out existing plant species and changing the ecological dynamics of its new residence.[34] Mellon and Rissler use the hypothetical example of a transgenic, salt-tolerant rice variety planted near coastal wetlands to illustrate the potential multiplier effects of a successful invasion. What would be the result, ask Mellon and Rissler, if the rice were to invade the nearby saltwater ecosystems, displacing indigenous salt-tolerant species?

> As the native populations decline, other organisms typically associated with them—such as algae, microorganisms, insects, other arthropods, amphibians, birds—might not be compatible with the invading rice. Different organisms, new to the salt-water marsh, might find homes in the new rice-dominated system.[35]

Molecular biologists working in the agricultural biotechnology industry argue that the addition of one or two transgenes into existing crops is not enough to confer weediness, and since the current technology limits the number of genes that can be successfully inserted into a plant, there's little cause for concern that transgenic crops might become weeds. However, a spate of recent studies on weediness belie the oft-heard claim of industry biologists that the likelihood of a transgenic crop becoming a weed is slim or nonexistent.

It is also argued by some that crops are so "non-weedy" that it is unlikely they could ever become successful weeds, even with the addition of traits designed to enhance their competitiveness. While some crops like corn have been so thoroughly weakened in the process of domestication that it is highly unlikely they could become successful weeds, Mellon, Rissler, and other scientists point out that many other crops retain weedy traits and are quite similar in their makeup to close weedy relatives, including alfalfa, barley, lettuce, potatoes, wheat, sorghum, broccoli, cauliflower, mustard, cabbage, and radishes. For all of the above reasons, Mellon and Rissler argue, "The possibility that engineering will convert crops into new weeds is a major risk of genetic engineering."[36]

A growing number of ecologists warn that an even bigger danger might lie in what is called "gene flow"—the transfer of transgenic genes from crops to weedy relatives by way of cross-pollination. This is just the kind of problem that worried some of the early agricultural biotechnology executives back in the 1980s when they pondered the liability and insurance issues surrounding the new science. If the resulting hybrids successfully reproduce with the weeds and pass the transgene along into subsequent generations, the transgene becomes part of the weed population.

Gene flows between crops and weedy relatives are naturally occurring and have been observed for more than a century by biologists. In California in the nineteenth century, a wild radish emerged as the result of hybridization between an "escaped" cultivated radish and an introduced weed known as jointed charlock. In Africa, a harmful weed, pearl millet, originated from the hybridization of millet and a wild relative, *Pennisetum americanum.* In France, a new weed evolved over the past several decades by the contamination of sugar beet with pollen from a wild Mediterranean subspecies. Wild rices have hybridized with cultivated rice, giving rise to wild weedy rice that often intermingles with the cultivated rice, creating untold problems for farmers. In Central America and Mexico, corn and its weedy relative teosinte have cross-pollinated, creating teosinte weeds which are often indistinguishable from the cultivated corn, and therefore often escape the weeding process.[37]

Researchers are concerned that transgenic genes for herbicide tolerance, and pest and viral resistance, might also escape and, through cross-pollination, insert themselves into the genomes of weedy relatives thereby creating weeds that are resistant to herbicides, pests, and viruses. Fears over the possibility of transgenic genes jumping to wild weedy relatives heightened in 1996 when a Danish research team, working under the auspices of Denmark's Environmental Science and Technology Department, observed the transfer of a transgene from a transgenic crop to the genome of a wild weedy relative—something critics of deliberate release experiments have warned of for years and biotech companies have dismissed as a remote or nonexistent possibility.

In the experiment, Dr. Thomas R. Mikkelsen and his team planted a transgenic oilseed rape plant containing a herbicide-resistant gene in a field near a close weedy relative, *Brassica campestris.* The plants cross-pollinated, creating hybrids. The hybrid seeds, in turn, were highly fertile and germinated into plants with the weedy features of the *Brassica campestris.* The research team reported that 42 percent of the second generation of the new weedy plants were found to be tolerant to the herbicide—clear evidence that the transgene had been successfully transferred from the oilseed rape to its weedy relative.[38]

The Danish study showed that transgenic genes inserted into crops could flow easily and rapidly into the wild, creating a new and virulent form of genetic pollution. Dr. G. A. de Zoeten, professor of plant pathology and plant molecular virology at Michigan State University, told *The New York Times* in an interview after the Danish study was published that "genes released into the environment eventually will escape, in essence creating a form of contamination."[39] Dr. de Zoeten added that the spread of genetic contamination through weeds could lead to unprecedented lawsuits, if the transgenes cross property lines into neighboring fields and lands.[40]

The prospect of weeds containing herbicide, pest, virus, and stress resistance spreading across agricultural fields and natural habitats is worrisome. Many of the transgenic genes being inserted into crops and readied for commercial introduction in countries around the world contain just the traits that are likely to provide a competitive advantage, if transferred to weeds in the wild.

Biotech industry spokespersons argue that the likelihood of transgenes flowing between crops to wild, weedy relatives is small, as most commercial crops are not grown near wild relatives. Recent studies, however, have shown that transgenes can migrate over much longer distances than previously thought. In one such study scientists planted genetically engineered potatoes containing an antibiotic-resistant gene. Ordinary potatoes were then planted at various distances from the transgenic crop. Thirty-five percent of the seeds collected from potatoes growing as far away as 1,100 meters from the transgenic potatoes contained the resistant gene.[41]

Crops like corn and soy have no weedy relatives in the United States and can be grown with little fear of spreading transgenes to the wild. Other crops, like squash, radish, carrots, and sunflowers, are grown near sexually compatible wild relatives and could spread genetic pollution into the wild. However, even in the case of crops like corn and soy, it is important to note that the chemical and agribusiness firms are preparing to market their transgenic seeds all over the world, virtually ensuring that in some regions transgenic crops will be grown near wild, weedy relatives, raising the prospect of contaminating centers of crop origin and diversity with this new form of genetic pollution.

In 1996, scientists from the Institute of Plant Sciences in Zurich reported that they had successfully transferred the *Bacillus thuringiensis* gene into India rice—the most common form of rice grown in the tropics—making the rice resistant to the yellow stem borer and striped stem borer. The International Rice Research Institute in Los Banos, Philippines, which financed the study, is now seeking approval to grow the transgenic rice in the Philippines. Some entomologists worry that the Bt rice, which is pollinated by wind, might spread to wild grasses that are close relatives, conferring pest resistance on the weeds and increasing the likelihood of creating resistant super bugs.[42]

The rapid globalization of commerce and the increased flow of international travel virtually guarantees that weeds contaminated with transgenes in one part of the world will eventually find their way to other regions, spreading genetic pollution over the planet. Transnational chemical and agribusiness companies project that within less than ten to fifteen years, all of the major crops will be genetically engineered to include herbicide-, pest-, virus-, bacteria-, fungus-, and stress-resistant genes. Millions of acres of agricultural land and commercial forest will be transformed in the most daring experiment ever undertaken to remake the biological world. Proponents of the new science, armed with powerful gene-splicing tools and precious little data on potential impacts, are charging into this new world of agricultural biotechnology, giddy over the potential benefits and confident that the risks are minimum or nonexistent. They may be right. But, what if they are wrong? What might be the consequences of unleashing herbicide-, pest-, virus-, bacteria-, fungus-, and stress-resistant genes into the biosphere?

Most molecular biologists and the biotechnology industry at large have all but dismissed the growing criticism of ecologists, whose recent studies suggest the likelihood of genetic pollution. Nonetheless, the uncontrollable spread of super weeds, the buildup of resistant strains of bacteria and new super insects, the creation of novel viruses, and the destabilization of whole ecosystems is no longer a minor consideration, the mere grumbling of a few disgruntled critics. To ignore the warnings is to place the biosphere and civilization in harm's way in the coming years. Pestilence, famine, and the spread of new kinds of diseases throughout the world might yet turn out to be the final act in the script being prepared for the Biotech Century.

## Designer Gene Weapons

The environmental threat posed by the release of genetically engineered organisms is likely to be compounded—perhaps dramatically—by the use of the new genetic techniques in the design of germ warfare agents. Breakthroughs in genetic engineering technology have renewed military interest in biological weapons and generated grave concern that an accidental or deliberate release of dangerous genetically engineered viruses, bacteria, and fungi could spread genetic pollution around the world, creating deadly pandemics that destroy plant, animal, and human life on a mass scale.

Current research in biotechnology parallels earlier research in the nuclear field in the 1940s and 1950s. The database developed for nuclear technology was applicable for both military and industrial purposes. Similarly, the database being developed for commercial genetic engineering in the fields of agriculture, animal husbandry, and medicine is potentially convertible to the development of a wide range of novel pathogens that can attack plant, animal, and human populations.

Biological warfare (BW) involves the use of living organisms for military purposes. Biological weapons can be viral, bacterial, fungal, rickettsial, and protozoan. Biological agents can mutate, reproduce, multiply, and spread over a large geographic terrain by wind, water, insect, animal, and human transmission. Once released, many biological pathogens are capable of developing viable niches and maintaining themselves in the environment indefinitely. Conventional biological agents include *Yersinia pestis* (plague), tularemia, rift valley fever, *Coxiella burnetii* (Q fever), eastern equine encephalitis, anthrax, and smallpox.

Biological weapons have never been widely used because of the danger and expense involved in processing and stockpiling large volumes of toxic materials and the difficulty in targeting the dissemination of biological agents. Advances in genetic engineering technologies over the past decade, however, have made biological warfare viable for the first time.

In a May 1986 report to the Committee on Appropriations of the U.S.

House of Representatives, the U.S. Department of Defense (DOD) pointed out that recombinant DNA and other genetic engineering technologies are finally making biological warfare an effective military option. Genetic engineers are cloning previously unattainable quantities of "traditional" pathogens. The technology can also be used to create novel pathogens never before seen. According to the report,

> . . . [Advances in biotechnology] permit the elaboration of a wide variety of "novel" warfare materials. . . . The novel agents represent the newly found ability to modify, improve, or produce large amounts of natural materials or organisms previously considered to be militarily insignificant due to problems such as availability, stability, infectivity, and producibility.[43]

The report goes on to say:

> Potent toxins which until now were available only in minute quantities, and only upon isolation from immense amounts of biological materials, can now be prepared in industrial quantities after a relatively short developmental period. This process consists of identifying genes, encoding for the desired molecule, and transferring the sequence to a receptive microorganism which then becomes capable of producing the substance. The recombinant organisms may then be cultured and grown at any desired scale. . . . Large quantities of compounds, previously available only in minute amounts, thus become available at relatively low costs.[44]

With recombinant DNA technology, it is now possible to develop "a nearly infinite variety of what might be termed 'designer agents.' "[45] The DOD report concludes that the new developments in genetic engineering technology make possible "the rapid exploitation of nature's resources for warfare purposes in ways not even imagined ten to fifteen years ago."[46] In August 1986, Douglas J. Feith, then Deputy Secretary of Defense, noted the near impossibility of defending against this newfound ability to genetically engineer biowarfare agents:

> It is now possible to synthesize BW agents tailored to military specifications. The technology that makes possible so-called "designer drugs" also makes possible designer BW. . . . It is [becoming] a simple matter to pro-

> duce new agents but a problem to develop antidotes. New agents can be produced in hours; antidotes may take years. To gauge the magnitude of the antidote problem, consider the many years and millions of dollars that have been invested, as yet without success, in developing a means of countering a single biological agent outside the BW field—the AIDS virus. Such an investment far surpasses the resources available for BW defense work.[47]

Recombinant DNA "designer" weapons can be created in many ways. The new technologies can be used to program genes into infectious microorganisms to increase their antibiotic resistance, virulence, and environmental stability. It is possible to insert lethal genes into harmless microorganisms, resulting in biological agents that the body recognizes as friendly and does not resist. It is even possible to insert genes into organisms that affect regulatory functions that control mood, behavior, and body temperature. Scientists say they may be able to clone selective toxins to eliminate specific racial or ethnic groups whose genotypical makeup predisposes them to certain disease patterns. Genetic engineering can also be used to destroy specific strains or species of agricultural plants or domestic animals, if the intent is to cripple the economy of a country.

The new genetic engineering technologies provide a versatile form of weaponry that can be used for a wide variety of military purposes, ranging from terrorism and counterinsurgency operations to large-scale warfare aimed at entire populations. Unlike nuclear technologies, genetically engineered organisms can be cheaply developed and produced, require far less scientific expertise, and can be effectively employed in many diverse settings.

Most governments claim that their biological warfare work is only defensive in nature. Yet it is widely acknowledged that it is virtually impossible to distinguish between defensive and offensive research in the field. Writing in the November 1983 edition of the *Bulletin of Atomic Scientists,* Robert L. Sinsheimer, a renowned biophysicist and former chancellor of the University of California at Santa Cruz, observed that because of the nature of this particular category of experimentation, there is no adequate way to properly distinguish between peaceful uses of deadly toxins and military uses.

The Stockholm International Peace Research Institute's exhaustive study on chemical and biological warfare concurs with Sinsheimer's assessment, concluding that "some common forms of vaccine production are very close technically to production of BW agents and so offer easy opportunities for conversion."[48]

Dr. Richard Goldstein, former professor of microbiology at Harvard Medical School, sums up the nature of the kinds of biological experiments currently being conducted by the DOD. Under the banner of defensive purposes, the DOD

> . . . can justify working with the super pathogens of the world—producing altered and more virulent strains, producing vaccines for protection of their troops against such agents . . . and likewise for the development of dispersal systems since DOD must be able to defend against any such dispersal system. Under this guise, what DOD ends up with is a new biological weapons system—a virulent organism, a vaccine against it, and a dispersal system. As you can gather from this, there is but a very thin line—if any—between such a defensive system (allowed by the conventions) and any prohibited offensive system.[49]

The Reagan administration was particularly concerned about the increasing interest in genetically engineered germ warfare agents and what it perceived as a "gene gap." In the fall of 1984, then Secretary of Defense Caspar Weinberger told members of Congress that he had obtained "new evidence that the Soviet Union has maintained its offensive biological warfare programs and that it is exploring genetic engineering to expand its program's scope." Weinberger went on to warn Congress, "It is essential and urgent that we develop and field adequate biological and toxin protection."[50] Convinced that the Soviets were violating the Biological Weapons Convention and widening the "gene gap" by launching a rigorous research and development program in genetic weaponry, the DOD announced intentions to respond, in turn, with its own ambitious "defensive" program.

Under the rubric of defensive research, the DOD launched a significant research and development effort in the 1980s. In 1981, the Pentagon budget for "defensive" biological warfare research was only $15.1 million. By 1986, the DOD budget had grown to $90 million.[51] The various branches of the armed services now work with virtually every major pathogen in the world, from exotic viral diseases such as hemorrhagic fevers, chikungunya, and dengue fever to newly discovered viruses such as AIDS. The DOD claims that most of the work is unclassified and intended to provide the military with defensive protection in the form of vaccines and antidotes.

Professional military observers are not sanguine about the prospect of keeping the genetics revolution out of the hands of the war planners. As a tool of mass destruction, genetic weaponry rivals nuclear weaponry, and it can be developed at a fraction of the cost. These two factors alone make genetic technology the ideal weapon of the future.

The recent revelation that Iraq had stockpiled massive amounts of germ warfare agents and was preparing to use them during the Persian Gulf War has renewed Pentagon interest in defensive research to counter the prospect of an escalating biological arms race. Saddam Hussein's government had prepared what it called the "great equalizer," an arsenal of twenty-five missile warheads carrying more than 11,000 pounds of biological agents, including deadly botulism poison and anthrax germs. An additional 33,000 pounds of germ agents were placed in bombs to be dropped from military aircraft. Had the germ warfare agents been deployed, the results would have been as catastrophic as those visited on Hiroshima and Nagasaki with the dropping of the atomic bombs in 1945. To get a sense of the potential damage that could have been inflicted, compare the Iraqi arsenal with a study conducted by the U.S. Office of Technology Assessment in 1993 that found that the release of just 220 pounds of anthrax spores from an airplane over Washington, D.C., could kill as many as 3 million people.[52] The Iraqi Scud missiles were filled with twice that amount of deadly anthrax. It was later reported that Saddam Hussein did not unleash the germ warfare agents because of a warning passed on to him by Secretary of State James Baker that any such effort would be met with "extreme measures," which meant the potential detonation of nuclear weapons over Baghdad.[53]

Iraq is not alone in its interest in developing a new generation of biological weapons. In a 1995 study, the Central Intelligence Agency (CIA) reported that seventeen countries were suspected of researching and stockpiling germ warfare agents. The nations include Iraq, Iran, Libya, Syria, North Korea, Taiwan, Israel, Egypt, Vietnam, Laos, Cuba, Bulgaria, India, South Korea, South Africa, China, and Russia.[54]

As knowledge of gene splicing becomes more sophisticated and accessible, it is likely that the next generation will be caught up in a deadly new biological arms race. The increasing experimentation with designer gene weapons in laboratories across the world, both for offensive purposes and defensive research, increases the likelihood of accidental releases. No laboratory, however contained and secure, is failsafe. Natural disasters such as floods and fires, and security breaches are possible and unpreventable. It is equally likely that terrorists and outlaws will turn to the new genetic weapons to spread fear and chaos as they seek to have their demands met by society.

In this century, modern science reached its apex with the splitting of the atom, followed shortly thereafter by the discovery of the DNA double helix. The first discovery led immediately to the development of the atomic bomb, leaving humanity to ponder, for the first time in history, the prospect of an end to its own future on Earth. Now, a growing number of military observers are wondering if the other great scientific breakthrough of our time will soon be used in a comparable manner, posing a similar threat to our very existence as a species.

## Animal Suffering

While some public attention has been focused, of late, on agricultural pollution and the potential effects of an accidental or deliberate release of deadly toxins and pathogens in biological warfare experiments, far less attention has been given to the impacts of genetic pollution on animal health, despite reports of a marked increase in animal suffering in transgenic research. Thousands of transgenic, chimeric, and cloned animals, from pigs to primates, are being experimented on in laboratories around the world in an effort to improve livestock, create more efficient ways of producing drugs and chemicals, and find cures for human diseases. Inserting foreign genes into the genetic code of an animal can trigger multiple reactions and result in unprecedented suffering for the creature.

Dr. Gill Langley, a fellow of the Royal Society of Medicine in the United Kingdom, lists several problem areas in genetic experimentation that can result in increased animal suffering. First, the insertion of a gene into an animal's chromosome is random. It is not uncommon, says Langley, for several or even hundreds of copies of a gene to be inserted in a row in a single location on a chromosome. Whether the transgene is expressed and how it is expressed is equally problematic and depends on the promoters, enhancers, and silencer genes in the host organism.

Second, if the transgene disrupts the host animal's natural genes, an insertional mutation can result. Langley cites the example of a transgene composed of a fruit fly gene and a viral gene (thymidine kinase) that was introduced into mouse embryos. Some of the newborn mice had extreme abnormalities including a loss of hind limbs, facial clefts, and massive brain defects.

Third, the transmission of transgenes from the founder animal to its offspring often fails to breed true, requiring hundreds of repeat experiments on additional animals to successfully develop the desired line. In addition, it is not uncommon for a transgene to be chemically altered in the offspring, producing totally unanticipated effects in the animals.

Fourth, transgenes may produce their product equally throughout an organ, but still have varying effects on different parts of the organ. In one experiment, transgenic mice were developed containing a cancer gene (from the SV40 virus) and a promoter from the atrial natriuretic factor gene designed to activate the cancer gene equally in both the upper and lower chambers of the heart. After birth, however, scientists were surprised to find the right upper chamber of the animal's heart growing uncontrollably, to hundreds of times its normal size, eventually overwhelming the rest of the organ. The left upper chamber, however, was unaffected by the transgene and grew normally in the transgenic mice. The researchers speculated that differences in the growth-factor levels in the two chambers may have played a role in the asymmetric response. The larger lesson is that the complex and multiple interactions between the inserted transgene and the chemical activity of the host animal are, for the most part, unknowable and unpredictable and can result in all sorts of novel and even bizarre pathologies in the creature.[55]

The public was exposed to the cruel suffering that can result from transgenic animal experiments in a rare television film clip aired on the national news several years ago. Scientists at the USDA research center in Beltsville, Maryland, had micro-injected a human growth hormone gene into the genetic code of pig embryos. The goal was to produce pigs that would grow larger and faster and generate increased profits for the livestock industry. The results, however, were quite different from what researchers had anticipated. Several of the animals showed gross abnormalities. One pig in the experiment was excessively hairy, arthritic, cross-eyed and lethargic. "The pitfall that we ran into," said Bob Wall, one of the two researchers on the USDA project, "was that we couldn't control the growth hormone gene in the manner we had hoped we could."[56] The human growth hormone gene USDA researchers had inserted either failed to trigger additional growth hormone or "kept the hormone faucet on constantly." Vern Pursel, the lead scientist conducting the transgenic experiments, reported that the muscles in the transgenic pigs degenerated and the animals became so weakened they could barely walk.[57]

Chastened by all of the adverse publicity over the failed experiment, USDA researchers abandoned the effort and turned their attention to inserting chicken genes into pig embryos to produce pigs with large shoulders. Pursel calls his newest transgenic "invention" "Arnie Schwarzenegger Pigs."[58]

Many of the transgenic animal experiments are designed to increase speed of growth, raise weight, and reduce fat. Critics argue that such experiments inevitably lead to increased stress on the animals, more health problems, and unnecessary suffering. Conventional animal husbandry practices bear witness to the cruel toll inflicted on animals in the interest of increasing profits. For example, the modern broiler chicken has been bred to grow to maturity in less than seven weeks and weigh more than five pounds at the time of slaughter. The animal's legs cannot hold its body weight and, as a result, it suffers from painful leg and foot deformities. Now, genetic engineers are seeking a patent in the European Patent Office for a transgenic chicken that contains a bovine (cow) hormone gene. The new transgenic chicken is designed to grow to maturity even faster, with leaner meat and earlier sperm production in males, virtually assuring more developmental abnormalities, greater stress, more risk of illness, and increased suffering.[59]

Another case in point is the much-heralded development of bovine growth hormone (BGH), also known as bovine somatotropin (BST). The genetically engineered product, produced by Monsanto under the trade name of Posilac, is designed to increase milk production in cows by as much as 20 percent and is currently marketed in the United States. While the genetically engineered hormone is now injected into cows in biweekly treatments, researchers are experimenting on inserting a transgenic growth hormone gene directly into the genetic code of the animals in the embryo stage. However, even with the more primitive injection process, the stress placed on the animals often leads to increased illness and suffering. The official U.S. Food and Drug Administration (FDA) package label that accompanies the product warns farmers that use of the gene-spliced hormone could result in a host of health-related problems. Reading from the label,

> [Using Posilac] may result in reduced pregnancy rates in injected cows. . . . Use of Posilac has also been associated with increases in cystic ovaries and disorders of the uterus during the treatment period. Cows injected with Posilac may have small decreases in gestation length and birth weight of calves and they may have increased twinning ratios. . . . Cows injected with Posilac are at an increased risk for clinical mastitis. . . . In some herds, use of Posilac has been associated with increases in somatic cell counts.[60]

Both the FDA and Monsanto acknowledge that cows treated with the genetically engineered hormone have a statistically greater chance of being afflicted with one or more of the illnesses listed on the official FDA warning label. But both parties argue that with "proper management practices" the health-related effects of the new drug should be minimized. Their assurances were not enough to calm angry dairy farmers across the country, many of whom were reporting increased health problems in their herds after administering the new genetically engineered drug. Anxious to document the mounting complaints, the National Farmers Union set up a special toll-free hotline in Wisconsin and was besieged by calls from dairy farmers whose animals had suffered and died from the stress-related impacts of the hormone injections.[61]

## The Rights of Other Creatures

"Context" is essential to any discussion of the underlying ethical considerations in developing transgenic animals. When the issue of creating transgenic animals is raised within a commercial context to excite would-be investors on Wall Street, molecular biologists often talk of the revolutionary potential of the new cutting-edge technologies. They boast of the ability to bypass millions of years of evolution and thousands of years of classical breeding and create wholly new bio-industrial designed creatures with unlimited commercial utility. When, however, the same transgenic animal experimentation is questioned by environmentalists and animal rights advocates, molecular biologists retreat to a far more conservative position, arguing that their efforts amount to little more than an extension of conventional breeding techniques.

The question of whether the creation of transgenic animals is fundamentally different or merely an extension of conventional breeding practices is critical to the debate over the legitimacy of the research. The issue was publicly raised in a frank and animated exchange of views among the nation's leading molecular biologists and critics in the spring of 1985 at an official meeting of the Recombinant DNA Advisory Committee (RAC) of the National Institutes of Health in Bethesda, Maryland. The Foundation on Economic Trends (FET) had formally requested that the NIH temporarily suspend funding certain transgenic animal experiments pending a review of the ethical implications and consequences of the research. The discussion around the NIH conference table that day centered on two related questions: Is transgenic animal research qualitatively new and therefore worthy of extended ethical deliberation and, if so, is the creation of transgenic animals morally permissible or a violation of the "inherent nature" or "beingness" of our fellow creatures.

Molecular biologists from around the country responded to the challenge raised by the FET petition. On the first issue, Dr. Cornelius Van Dop of the Johns Hopkins Hospital in Baltimore expressed the views of virtually every biologist involved in the research when he wrote:

> The selective breeding of animals directed to amplifying or eliminating certain traits has been a human activity since the first mammal was domesticated during prehistoric times. This selection for specific traits (mutated genes) has irreversibly modified the gene pools of innumerable species for man's economic gain and whim. . . . Current bioengineering technology stands at the threshold of being able to selectively modify one gene at a time and thereby reduce dependence on selective breeding for

> altering certain traits. The selective introduction of foreign genes into germ lines is thus a logical extension of animal husbandry.[62]

Despite what many molecular biologists claim, transgenic animals are more than the result of a sophisticated extension of traditional breeding practices. Classical breeders could never have produced giant super mice containing human growth hormone genes that grow to twice the size of normal mice. Nor could classical breeders ever have crossed a sheep and a goat, two totally unrelated species, creating a chimeric animal—the "geep"—that is half goat and half sheep. Nor could conventional breeding technologies hope to create the now famed Dolly—a cloned sheep and the first mammal in all of history born of the process of replication rather than conception.

Transgenic animals are a radical departure from both evolutionary history and classical breeding practices. Never before in history have scientists had the tools to bypass species boundaries altogether and create novel creatures by combining genetic information from across the expanse of the biological kingdom. One of the participants in the NIH debate hinted at the real excitement that underlies so much of the current research, but that is rarely expressed openly for fear of alarming the public. For the first time, molecular biologists believe they are within grasp of capturing control over the evolutionary process itself, of dictating the terms, albeit in a very primitive fashion, of nature's developmental journey. Dr. David Martin, Jr., of the School of Medicine of the University of California at San Francisco, opined,

> We humans are participating in the process of evolution *per se.* By that I mean that our ability, acquired through evolution, to manipulate genomes by selective breeding, and more recently by recombinant DNA technology, is an integral component of evolution itself and is not, as has been claimed in the past, "tinkering with evolution." Instead, *it is* evolution.[63]

If these new transgenic tools allow for a degree of authorship over other creatures, far in advance of the kinds of manipulations we have been able to exercise in the past, then the question becomes one of whether or not our fellow creatures ought appropriately to be the subject of wholesale reconfiguration along new developmental lines. In other words, is there an argument to be made for honoring the "speciesness" or "beingness" of the many animal species that exist on Earth?

The notion of intrinsic value and speciesness were soundly rejected by the molecular biologists, who believe such terms belong to the world of "mysticism" and have no place in the discussion of scientific issues. Dr. Maxine Singer, of the NIH, remarked, "History, from Galileo through Lysenko, teaches us that mysticism can never yield rational and wise public policy in scientific matters. . . ."[64]

It's not difficult to understand why molecular biologists are so dismissive of the idea of "speciesness." Crossing species boundaries is the essence of the new biotechnology revolution. To acknowledge even the remotest possibility of a moral, ethical, or philosophical case for species preservation is to question the very nature of genetic engineering technology.

This doctrinaire denial of "speciesness" puts many of the molecular biologists at odds with a growing number of environmental scientists, who have come to view the preservation of species as both an environmental and moral imperative. The blanket assertion that animals have no intrinsic value also places the molecular biologists well outside the mainstream of public opinion. According to an Associated Press poll conducted in 1995, 67 percent of Americans agree somewhat or strongly that animals have the right to pursue their own natural and essential interests and that an animal's right to live free of suffering should be as important as a person's right to live free of suffering. An earlier poll commissioned by the USDA found that a majority of Americans oppose the transfer of genes between unrelated species on ethical grounds.[65]

While the majority of the public believes that regard for the rights of other animals is important and ought to be respected, the scientists involved in the RAC debate felt otherwise, arguing that morality and ethics play no role in science. Dr. David Baltimore, a Nobel Laureate and former director of the Whitehead Institute of Biomedical Research in Cambridge, Massachusetts, said as much, declaring his unequivocal opposition to "writing into regulations statements about 'morally and ethically unacceptable' practices because these are subjective grounds and therefore provide no basis for discussion."[66]

Baltimore and his colleagues never harbored a moment of doubt that their perception of nature and species was informed by "objective truth," untainted and unaffected by "subjective" human values. But herein lies the nub of the problem. On what basis do we decide the worth, nature, and essence of species and individual organisms, including human beings, if it is not by way of the "subjective" values we hold? In asserting that animals have no beingness, no essential nature, no "speciesness" that ought to be regarded as of intrinsic worth, are not Baltimore and his colleagues expressing their own subjective values about how they conceive of life and nature? When molecular biologists choose to create transgenic animals, what values motivate their research? Certainly, notions of utility, efficiency, profitability, and even the more vague notion of "progress" are steeped in the values of the modern Enlightenment tradition and are every bit as subjective as the moral and ethical claims on behalf of defending and preserving the intrinsic worth and rights of our fellow creatures.

The question, then, raised by the new biotechnologies, and made more immediate and poignant by the sudden creation of clones, chimeras and other transgenic animals, is how should we regard our fellow species—and ultimately our own—in the coming Biotech Century? How do we value the many creatures that travel with us here on Earth? How do we see our relationship to them? The answer to that question will determine the kind of science we practice and the kind of world we inhabit in the next age of history.

## Human Health

Human beings may yet turn out to be the ultimate guinea pigs in the radical experiment to reseed the Earth with a laboratory-conceived second Genesis. The introduction of novel genetically engineered organisms raises a number of serious and even life-threatening human health issues. While mention has already been made of the potentially catastrophic consequences of an accidental or deliberate release of deadly genetically engineered biological warfare agents, even the seemingly mundane uses of genetic engineering in the production of foods poses a very real danger to human health.

The FDA announced in 1992 that special labeling for genetically engineered foods would not be required, touching off protests among food professionals, including the nation's leading chefs and many wholesalers and retailers. Critics are worried that the introduction of novel genes into conventional foods could trigger serious allergic reactions in people. With 2 percent of adults and 8 percent of children having allergic responses to commonly eaten foods, consumer advocates argue that all gene-spliced foods need to be properly labeled so that consumers can avoid health risks.[67] Their concerns were heightened in 1996 when *The New England Journal of Medicine* published a study showing that genetically engineered soybeans containing a gene from a Brazil nut could create an allergic reaction in people who were allergic to the nuts. Scientists at the University of Nebraska tested blood serum from nine subjects who are allergic to Brazil nuts against both an extract from genetically altered soybeans containing a gene from the Brazil nut and an extract from conventional soybeans. All of the serum reacted to the soybeans containing the Brazil nut gene and none reacted to the non-altered soybeans. The test result was unwelcome news for Pioneer Hi-Bred International, the Iowa-based seed company that hoped to market the new genetically engineered soy. The biotech industry had long dismissed critics who warned of the potential allergenic effects of introducing foreign genes into conventional food crops. The Nebraska study gave added weight to critics' concerns. The editorial board of *The New England Journal of Medicine* stated in an editorial that the study confirmed "that food allergens could indeed be transferred from one plant to another by transgenic manipulation."[68]

Many of the genes being transferred into the genetic code of food crops come from plants, microorganisms, and animals that have never before been part of the human diet. Worried by the findings of the Nebraska study, the *Journal* editors warned their readers that "the allergenic potential of these newly introduced microbial proteins is uncertain, unpredictable, and untestable."[69] Although the FDA said it would label any genetically engineered foods containing genes from common allergenic organisms, the agency fell well short of requiring across-the-board labeling, leaving the *Journal* editors to ask what protection consumers would have against novel genes from organisms that have never before been part of the human diet and that might be potential allergens. Concerned over the agency's seeming disregard for human health, the *Journal* editors concluded that FDA policy "would appear to favor industry over consumer protection."[70]

In the coming years, agrichemical and biotech companies plan on introducing hundreds, even thousands of genes into conventional food crops and food animals from bacteria, viruses, fungi, and non-food plants and animals—including human beings—raising the very real possibility of triggering new kinds of allergenic responses about which little is known and for which there exist no known treatments. Some of the allergies could prove serious and even life-threatening.

Consumers whose religious teachings proscribe a certain diet, as well as vegetarians, have also expressed concern over the FDA's decision not to require labeling. Jews and Muslims would not know, for example, whether their food contains a gene from pigs. Nor would vegetarians know if a vegetable they are eating contains animal genes. A 1997 public opinion survey commissioned by the transnational conglomerate Novartis found that 93 percent of the public thought that all biotech food should be labeled.[71]

Although the prospect of creating new food allergies has health professionals clearly worried, an even more serious potential health risk has recently attracted the attention of scientists and physicians. Some researchers are warning that the transplanting of genetically altered animal organs into humans could result in animal viruses crossing species boundaries, creating deadly new viral epidemics for which there are no cures. That concern increased in the mid-1990s in the wake of several baboon-to-human organ transplants. With biotech companies anxious to supply genetically modified animal organs to tens of thousands of patients in need of transplants in the coming years, virologists are warning of the unleashing of viral pandemics of "nightmare" proportions.

Once considered far-fetched by some within the medical community and only a remote possibility by others, researchers now acknowledge that the likelihood of deadly viruses breaking out from their host species and invading the human genome is a very real and dangerous prospect. The worldwide AIDS epidemic has alerted both virologists and the public to the devastating consequences that can follow from cross-species virus transfers. The HIV virus is suspected to have originated in the West African rain forests and spread to people by way of monkeys. Dr. Jon Allan of the Department of Virology and Immunology of the Southwest Foundation for

Biomedical Research in San Antonio, Texas, says, "Placing animal organs directly into humans with a concoction of immunosuppressive regimens, however, may have far greater consequences than any rainforest microbe."[72]

The public first became familiar with xenotransplants in 1984, when a fifteen-day-old baby was given a baboon heart to replace her own defective one. Baby Faye died twenty days later. In 1992, a thirty-five-year-old man received a baboon liver at the University of Pittsburgh School of Medicine and survived for two and a half months. In 1995, Jeff Getty, a resident of San Francisco infected with AIDS, received a bone marrow transplant from a baboon.[73] While the baboons used in these experiments were screened for the six known exogenously transmitted retroviruses and herpes viruses, researchers point out that the animals could be reservoirs for other, yet unknown, viruses that could slip by undetected in the harvesting and transplantation process.

Fear of using baboon and other monkey organs for xenotransplantation has led biotech companies to experiment with pig organs as a more suitable and potentially safer alternative. Researchers argue that swine, unlike monkeys, are raised in pathogen-free environments and are free of the kind of dangerous viruses that could take up residence in the human population. Their confidence was shaken in 1997, however, when scientists reported the discovery of a porcine endogenous retrovirus (PERV) that infected human cells in vitro, raising the possibility that other, undiscovered, pig retroviruses could jump species boundaries during transplantation and create disease in human patients. Still more frightening, says Allan, is that many retroviruses are blood-borne or sexually transmitted pathogens and, therefore, PERVs could be passed on through human contact with others and, like the AIDS virus, create an epidemic. Allan says that the results of these latest findings on porcine retroviruses "should compel public health agencies in the United States and elsewhere to regard animal-to-human transplantation as a melting pot for retroviruses that might result in recombinant viruses with altered pathogenicities."[74] Even the editors of *The Economist,* normally the first to champion new technologies with vast market potential, urged caution on xenotransplants. "Simple prudence, not alarmism, suggests that it is not yet time to realize the surgeon's dream of an endless supply of organs from beasts."[75]

The federal government, however, bowing to pressure from the biotech industry and the medical community, threw caution to the wind in 1996, permitting the use of both pig and baboon organs for transplantation. Incredibly, the FDA and the Centers for Disease Control agreed to let local surgeons and institutional review boards police their own xenotransplantation practices. The laissez-faire policy being pursued in the United States contrasts sharply with that of the United Kingdom, where the government has barred all xenotransplants pending further study of the risks of spreading animal viruses into the human population.[76]

Perhaps the most revealing indication of the power of the transplantation lobby in the United States is to be found in the findings and recommendations of the government's Institute of Medicine (IOM). An IOM panel stated that "there is every reason to believe that the potential for transmission of infectious agents . . . from animals to human transplant recipients is real." Even so, the panel concluded, "the potential benefits of xenotransplants are great enough to justify this risk."[77]

## Depleting the Gene Pool

Ironically, all of the many efforts to create a bio-industrial future may eventually come to naught because of a massive catch-22 that lies at the heart of the new technology revolution.

On the one hand, the success of the biotech revolution is wholly dependent on access to a rich reservoir of genes to create new characteristics and properties in crops and animals grown for food, fiber, and energy, and products used for pharmaceutical and medical purposes. Genes containing novel and useful traits that can be manipulated, transformed, and inserted into organisms destined for the commercial market come from either the wild, from landraces (traditional crops) and domesticated animal breeds, and from human beings. Notwithstanding its awesome potential to transform nature into commercially marketable commodities, the biotech industry still remains utterly dependent upon nature's seed stock—germplasm—for its raw resources. At present it is impossible to create a "useful" new gene in the laboratory. In this sense, biotechnology remains an extractive industry. It can "mine" genetic material, but cannot create it *de novo.*

On the other hand, the very practice of biotechnology—gene splicing, tissue culture, clonal propagation, and monoculturing—is likely to result in increased genetic uniformity, a narrowing of the gene pool, and loss of the very genetic diversity that is so essential to guaranteeing the success of the biotech industry in the future. Biologist Peter Raven, director of the Missouri Botanical Gardens in St. Louis, put it best in an interview with the editors of *Genetic Engineering News* several years ago. He noted, "Both the progress and profitability [of the biotech industry] depend on their ability to understand and manipulate biological diversity."[78]

The importance of biological diversity has been most apparent in addressing the many devastating blights that have ravaged agriculture for more than 150 years. Blight, in modern agriculture, is the result of planting pure line strains—monocultures—making the crops potentially vulnerable to particular viral, bacterial, and fungal diseases. In each case, farmers and consumers were ultimately saved by the introduction of new crops with genes that were found to be resistant to the infestation. The new plants were discovered either in the wild or were landraces whose genetic composition made them resistant to the blight spreading through the fields.

The first and perhaps best known example of modern blight occurred in Ireland in the 1840s. Potatoes were discovered in the New World and quickly found their way back to Europe. The potato soon became a staple in the Irish diet. In 1845, a mysterious blight attacked the potato crop and persisted for five years, spreading famine across Ireland. More than a million people died and many others migrated to North America to escape the famine. The Irish potatoes were descendants of a genetically limited stock that turned out to be highly vulnerable to the blight. Eventually researchers were able to find new strains of the potato in the Andes and Mexico, where the plant originated, that were resistant to the blight.[79]

In the 1870s coffee rust devastated the coffee crops in India and Ceylon (Sri Lanka), and in 1904 stem rust attacked and crippled the wheat crop in the United States. In 1943, brown spot disease destroyed much of India's rice crop, resulting in famine in Bengal. In the 1970s corn blight attacked the corn crop in the American South. In each instance, modern agriculture's propensity to plant pure strains or monocultures in the fields made them vulnerable to wholesale destruction. New strains with resistant traits were eventually found to replace the vulnerable strains, again planted in monocultures, again setting up the condition for future blights.[80]

A growing number of scientists and observers are becoming worried that the loss of genetic diversity on Earth is narrowing the prospects for providing new food, pharmaceuticals, and fiber for the human race and are beginning to urge governments to protect and preserve the "green gold." The United Nations Food and Agricultural Organization (FAO) estimates that some forty thousand valuable plant species will become extinct by the middle of the twenty-first century. Edouard Saouma, FAO Director General, warned that "their loss constitutes a grave threat to our world food security."[81] In his book *The Last Harvest,* Paul Raeburn, the science editor for *Business Week,* penetrates to the heart of the problem. He writes,

> Scientists can accomplish remarkable feats in manipulating molecules and cells, but they are utterly incapable of re-creating even the simplest forms of life in test tubes. Germplasm provides our lifeline into the future. No breakthrough in fundamental research can compensate for the loss of the genetic material crop breeders depend on.[82]

Raeburn's and others' concerns are coming in the midst of unprecedented species losses resulting from growing human pressures on ecosystems. For example, overgrazing, logging, and the expansion of human settlements in the Andes are pushing wild potatoes to the edge of extinction. In the coastal region of Peru, overgrazing of goats is threatening the extinction of wild tomato species. Deforestation in Central America, to make room for oil exploration by Phillips Petroleum and Texaco, resulted in the wholesale loss of cocoa varieties in that region of the world.[83] Donald Falk, former director of the Center for Plant Conservation at the Missouri Botanical Gardens in St. Louis, says that between three thousand and five thousand of the twenty-five thousand plants native to the United States are nearing the point of extinction.[84]

Biologist E. O. Wilson, of Harvard University, estimates that thousands of plant species found in Central and South America will likely become extinct within the next hundred years. Included in the list will be many of the wild relatives of tomatoes, corn, peanuts, beans, peppers, squash, and cocoa, all of which originated in that part of the world. Wilson predicts, "If destruction of the rain forest continues at the present rate to the year 2022, half of the remaining rain forest will be gone. The total extinction of species this will cause will be somewhere between ten percent and twenty-two percent." He says these catastrophic losses represent between 5 and 10 percent of the species currently existing in the world. Wilson calculates that we are currently losing twenty-seven thousand plant and animal species each year. That means that seventy-four species are being lost every day. Wilson concludes, "We are in the midst of one of the great extinction spasms of geological history."[85]

The commercial value of these losses is potentially enormous. In their book *The First Resource,* Christine and Robert Prescott-Allen calculate that one out of every twenty-two dollars generated in the United States—4.5 percent of the gross domestic product—is derived from wild species. The authors argue that the commercial value of wild germplasm has been largely ignored in recent years by economists and business and policy leaders who prefer to focus their attention more on the commercial value of fossil fuels and other non-renewable resources. As a result, most people in the industrial countries regard wildlife less as a commercial asset and more as an "aesthetic, emotional, and recreational" resource.[86] That's likely to change as we move from an industrial to a biotech economy and from fossil fuels to genes as our primary natural resource.

The loss of genetic diversity is compounded by modern farming practices that continue to emphasize monoculturing over mixed cropping methods. Agribusiness and chemical companies are always on the lookout for the "perfect" product, a plant strain that will grow quickly, be resistant to disease, and be easy to pick and transport to market. Market forces in both the developed and developing world have conspired to force farmers to switch from the growing of landraces to the growing of high-performance monocultures. The abandonment of the enormous number of traditional varieties in favor of the new strains has seriously undermined genetic diversity, creating over-reliance on a dwindling number of plant genomes.

Genetic erosion is already well advanced in most countries. The U.S. soy crop, which accounts for 75 percent of the world's soy, is a monoculture that can be traced back to only six plants brought over from China. The Rural Advancement Foundation International (RAFI) reports that of the seventy-five kinds of vegetables grown in the United States, 97 percent of all the varieties have become extinct in less than eighty years. According to the RAFI study, of the 7,098 apple varieties grown in the United States between 1804 and 1905, 6,121 or 86.2 percent have since become extinct. Of the 2,683 pear varieties in use in the last century, 2,354 or 87.77 percent are now extinct.[87] The grim statistics are repeated for every food crop. Martin Teitel, in his book *Rain Forest in Your Kitchen,* points out that in the United States just ten varieties of wheat account for most of the domestic harvest, while only six varieties of corn make up more than 71 percent of the yearly crop.[88] In India, farmers grew more than thirty thousand traditional varieties of rice just fifty years ago. Now, ten modern varieties account for more than 75 percent of the rice grown in that country.[89] Garrison Wilkes, professor of botany at the University of Massachusetts, says that the spread of modern agricultural practices is quickly destroying the genetic resources upon which it is built and likens the situation to "taking stones from the foundation to repair the roof."[90]

Agricultural biotechnology will only intensify the practice of monoculturing, as did the Green Revolution when it was introduced more than thirty years ago. Like its predecessor, the goal of the biotech revolution is to create superior varieties that can be planted as monocultures in agricultural regions all over the world. A handful of agribusiness and chemical companies are staking out the new biotech turf, each aggressively marketing their own patented brands of "super seeds"—and soon transgenic farm animals as well. The new transgenic crops and animals are designed to grow faster, produce greater yield, and withstand more varied environmental and weather-related stresses. Their cost effectiveness, in the short run, is likely to guarantee them a robust market. In an industry where profit margins are notoriously low, farmers will likely jump at the opportunity of saving a few dollars per acre and a few cents per pound by shifting quickly to the new transgenic crops and animals. However, the switch to a handful of "the best" patented transgenic seeds and animals will likely further erode the genetic pool as farmers abandon the growing of traditional varieties and breeds in favor of the commercially more competitive transgenic products.

The continued diminution of the world's remaining agricultural germplasm is being hurried along by the introduction of more sophisticated propagation methods, including cloning and tissue culture. With these new technologies, it is possible to mass-produce identical copies of an original genotype, each indistinguishable from the other. Scientists can now take a single gram of callus (a cluster of undifferentiated plant cells) and produce more than ten million plant embryos in a cultivation tank in six months, bypassing growing seasons, farmers, and the vagaries and uncertainties that accompany traditional planting of seeds in the fields.[91]

Increasingly sophisticated techniques for mass-producing plant clones have already remade agricultural practices. The recent cloning of a mammal is likely to reshape animal husbandry in similar ways in the coming years. In the frenzied public discussion following the announcement of Dolly's birth, scant commentary touched on the underlying significance of the story. The cloning of animals introduces replication into the biological process and with it the specter of using industrial engineering principles to both customize and mass-produce identical copies of living organisms. In this regard, Dolly's birth is a breakthrough event in the emerging Biotechnical Age.

Agribusiness, pharmaceutical, and biotech companies are already seeking patents on "superior" animal genotypes that can be replicated and mass-produced for use as chemical factories, for organ transplantation, and for meat consumption. Industry leaders hope to create brand-name identification for their cloned animals and plan on spending large sums of money on marketing their Dollys and Pollys. With genetic diversity already seriously compromised in domestic herds as a result of years of monoculturing, critics worry that cloning will all but eliminate any remaining genetic variation and force reliance on a dangerously diminished genetic pool. We may be facing a new era in animal husbandry where only a handful of brand-name genotypes exist for cows, sheep, pigs, and other animal species, the rest being eliminated by the forces of the market.

Transgenic crops threaten to drain the world's genetic reservoirs in still other ways. Unlike classical breeding, in which resistance has been built into the plants over long periods of time, and often involves hundreds of genes, the genetic engineers rely on introducing only one or two resistant traits in hopes of warding off any potential environmental assault. The old-fashioned approach is far more time consuming and complicated and well beyond the technical capabilities of the best scientists in the world. In short, the new transgenic technologies are quite primitive compared to nature's own processes, a point often missed in the hoopla created around the new genre of gene-splicing techniques. All of the transgenic introductions amount to little more than a quick technological fix, a short-lived set of gerrymandered solutions that virtually guarantee an even quicker and more virulent response by their natural enemies. By relying on only "one gene" resistance, the genetic engineers make it far easier for insects, viruses, and fungi to triumph, making their transgenes useless in a very short period of time. In the meantime, the traditional varieties that may contain hundreds of genes working together in myriad ways to ward off diseases are abandoned to extinction to make way for the new transgenic "super plants." This means that enough genetic ammunition might not be available, in the future, in the form of additional resistant genes, to continue to provide defenses against continued waves of ever more resistant weeds, insects, viruses and the like. Longtime critics of the new agricultural biotechnology Cary Fowler and Pat Mooney write,

> In the process of going after the single gene for resistance, the gene-complex—the whole set of genes that can provide stable resistance in a landrace—is often ignored, and sometimes destroyed, despite its 'representing all the plant breeding work carried out by nature over thousands of years.' . . . The loss of crop genetic resources through extinction and the squandering of the remaining resources though one-gene resistance breeding reduce the odds of our being able to counter pests and diseases successfully in the future.[92]

Transgenic crops pose an even more direct threat to the world's remaining centers of crop diversity. These centers are the regions that contain both wild relatives and landraces and are the reservoirs for providing new genetic material for purposes of breeding. There is growing concern that the large-scale introduction of transgenic crops could contaminate the world's remaining centers of crop diversity. Gene flow from transgenic plants to landraces is likely inevitable in the wake of ambitious plans by the biotech industry to aggressively market their new "super seeds" in every agricultural region of the world. It will probably be impossible to shield the few remaining centers of crop diversity from the increasing encroachment of transgenic crops.

National governments have shown little or no inclination to address the many concerns being raised by traditional plant breeders, farmers, and environmental critics over this potentially dangerous situation. Much of the reason there is so little concern being expressed about these worrisome environmental issues can be attributed to the powerful commercial forces at work. Global chemical and pharmaceutical giants are moving quickly to consolidate their control over the world's remaining germplasm reserves. As mentioned earlier, seed companies are being bought up by chemical companies who hope to own the "green gold" of the Biotech Century.[93] For the chemical companies, being able to control the distribution of patented seeds assures them virtual hegemony over much of global agriculture. Pharmaceutical firms, aware that more than a quarter of all new prescription drugs are derived from plant material, are anxious to control as much of the world's plant stocks as possible to ensure their dominance in the global drug market. Although these companies should be concerned about the long-term erosion of genetic resources, their commercial horizon rarely extends beyond the immediacy of the market and the short-term profit potential of selling a limited number of high-visibility "brand" seeds, animal genotypes, and patented drugs.

The commercial enclosure of the world's seeds—once the common inheritance of all humankind—in little less than one century, while hardly given more than a passing notice in the media, is, nonetheless, one of the more important developments of modern times. Just a century ago, hundreds of millions of farmers, scattered across the planet, controlled their own seed stocks, trading them freely among neighbors and friends. Today, much of the seed stock has been brought up, engineered, and patented by global companies and kept in the form of intellectual property. Farmers wishing to plant for future harvests are increasingly reliant on access to these same companies, to whom they have to pay a fee for use of what was a commonly held good a short time ago. For their part, the chemical and pharmaceutical companies have little desire to champion the interests of small peasant and independent farmers around the world who still grow traditional landraces, passing on their heirloom crops from one generation to another. The independent farmer, growing traditional varieties, is seen less as a curator of potentially valuable resources and more as a potential market for the new patented seeds. The biotech corporations seek his business and make every effort to sell him their brand of seeds. By focusing on short-term market priorities, the biotech industry threatens to destroy the very genetic heirlooms that might one day be worth their weight in gold as a new line of defense against a new resistant disease or super bug.

The reseeding of the planet with a laboratory-conceived second Genesis is likely to enjoy some enviable short-term market successes, only to ultimately fail at the hands of an unpredictable and noncompliant nature. While the genetic technologies we've invented to recolonize the biology of the planet are formidable, our utter lack of knowledge of the intricate workings of the biosphere we're experimenting on poses an even more formidable constraint. The introduction of new genetic-engineering tools and the opening up of global commerce allow an emerging "life industry" to "reinvent" nature and manage it on a worldwide scale. The new colonization, however, is without a compass. There is no predictive ecology to help guide this journey and likely never will be, as nature is far too alive, complex, and variable to ever be predictably modeled by scientists. We may, in the end, find ourselves lost and cast adrift in this artificial new world we're creating for ourselves in the Biotech Century.

# Four A Eugenic Civilization

When Aldous Huxley wrote his dystopian novel *Brave New World* in 1932, neither he nor his contemporaries could have imagined that by the end of the twentieth century the scientific insights and technological know-how would be in place to make real his vision of a eugenic civilization. The mapping of the human genome, the increasing ability to screen for genetic diseases and disorders, the new reproductive technologies, and the new techniques for human genetic manipulation comprise the fourth strand of the operating matrix of the Biotech Century and establish the technological foundation for a commercial eugenics civilization. Human gene screening and therapy raise the very real possibility that, for the first time in history, we might be able to reengineer the genetic blueprints of our own species and begin to redirect the future course of our biological evolution on Earth. The prospect of creating a new eugenic man and woman is no longer just the dream of wild-eyed political demagogues but rather a soon-to-be-available consumer option and a potentially lucrative commercial market.

Genetic engineering technologies are, by their very nature, eugenics tools. Because the technology is inseparably linked to eugenics ideas, no thoughtful discussion of the new technology revolution can occur without raising eugenics issues. The term "eugenics" was conceived by Sir Francis Galton, Charles Darwin's cousin, in the nineteenth century and is generally divided along two lines. Negative eugenics involves the systematic elimination of so-called undesirable biological traits. Positive eugenics is concerned with the use of selective breeding to "improve" the characteristics of an organism or species.

Eugenics found its first real home in America at the turn of the century. The rediscovery of Mendel's laws spurred renewed interest in heredity within the scientific community. The new discoveries in heredity were used by geneticists and social reformers to kindle a massive eugenics movement in the popular culture. By the time that movement had finally run its course during the Great Depression, American society was awash with eugenics dogma. Many Americans came to believe that blood ties and heredity were far more important in shaping individual behavior and in determining the status of various ethnic and racial groups than economic, social, or cultural determinants.

The history of the American eugenics movement needs to be publicly aired, especially in light of the many new scientific discoveries and inventions that now make possible the kind of eugenics society that earlier eugenics reformers only could have dreamed of achieving. At a time when so many mainstream politicians, scientists, academicians and editorial writers discount the likelihood of a eugenics movement emerging in the twenty-first century, America's eugenics past is a sobering reminder that "it can happen here."

## America's Eugenic Past

> Some day we will realize that the prime duty, the inescapable duty of the good citizen of the right type is to leave his or her blood behind him in the world; and that we have no business to permit the perpetuation of citizens of the wrong type. The great problem of civilization is to secure a relative increase of the valuable as compared with the less valuable or noxious elements in the population. . . . The problem cannot be met unless we give full consideration to the immense influence of heredity. . . . I wish very much that the wrong people could be prevented entirely from breeding; and when the evil nature of these people is sufficiently flagrant, this should be done. Criminals should be sterilized and feeble-minded persons forbidden to leave offspring behind them . . . the emphasis should be laid on getting desirable people to breed.[1]

This quote could have come from the lips of countless political functionaries at party rallies and meetings throughout Nazi Germany in the 1930s. But it didn't. It was uttered by the twenty-sixth President of the United States, Theodore Roosevelt, and it represented the "enlightened" view of millions of Americans caught up in an ideological movement that has been virtually written out of American history books.

From the turn of the century until the Great Depression, eugenics was embraced by much of America's intellectual elite as the cure-all for the economic inequities and social ills that were then threatening the fabric of American life. It took hold at a time when reformers were becoming increasingly disheartened by their seeming inability to deal effectively with the escalating problems of poverty, crime, and social unrest.

The eugenics movement was spawned in the 1890s in the wake of the first massive immigration wave, which brought with it a mushrooming growth of city slums and militant union-organizing drives. It reached its peak in the chilling isolationist atmosphere following World War I, which produced the first great Red scare in America. During this time, America's old-line ruling families combined forces with middle-class academics and professionals in an active alliance to promote the notion of a eugenics policy for the United States. America's white Anglo-Saxon Protestant elite was becoming increasingly paranoid over its loss of control of the economic and political machinery of the country. For the first time, WASP hegemony was being vigorously challenged by the Irish, the Jews, the Italians, and other immigrant groups demanding a piece of the American Dream.

At the same time, professionals and academics were desperately looking for a way to explain their failures in the area of social and economic reform. Both groups found the answer in eugenics. Its attractiveness was irresistible. First, its premise—that heredity, not environment, determined the behavior of people in society—allowed the reformers the excuse they needed to place the blame on the masses for the wrongs that beset society. The upper class saw in eugenics a philosophical rationale that they could seize on in order to protect their claims to power. More important still, at a time when science was being heralded as the linchpin of American greatness and a road map to its manifest destiny, eugenics offered a scientific explanation for social and economic problems and a scientific approach to their solution. Historian Mark H. Haller points out that the eugenics movement became enormously powerful and influential precisely because it appealed to the "best people."

It was the "best people" who overnight turned eugenics into a form of secular evangelism. They preached their newfound creed in university lecture halls, before professional conventions, and on political platforms from one end of the country to the other. And the message was always the same:

America's salvation hinged on its resolve to eliminate the biologically inferior types and breed a superior stock of men and women.

Leading American geneticists were responsible for spearheading much of the early eugenics movement. According to Kenneth Ludmerer in his seminal work, *Genetics and American Society*, nearly half the geneticists in the country became involved in the eugenics movement in one way or another. Many were "alarmed by what they considered to be a decline in the hereditary quality of the American people."[2] Scientists became active in leadership roles in the eugenics cause in the hope "that they could help reverse the trend."[3] Michael F. Guyer of the University of Wisconsin boldly proclaimed that "the fate of our civilization hangs on the issue."[4] The famed geneticist Edward G. Conklin dispassionately observed "that although our human stock includes some of the most intelligent, moral and progressive people in the world, it includes a disproportionately large number of the worst human types."[5] Professor H. S. Jennings of Johns Hopkins University informed the American public,

> The troubles of the world and the remedy of these troubles lie fundamentally in the diverse constitutions of human beings. Laws, customs, education, material surroundings are the creations of men and reflect their fundamental nature. To attempt to correct these things is merely to treat specific symptoms. To go to the root of the troubles, a better breed of men must be produced, one that shall not contain the inferior types. When a better breed has taken over the business of the world, laws, customs, education, material conditions will take care of themselves.[6]

In 1906 the American Breeders Association set up the first functioning Committee on Eugenics. Its stated purpose was to investigate and report on heredity in the human race and to emphasize the value of superior blood and the menace to society of inferior blood.[7] The Committee's membership included such distinguished Americans as horticulturist Luther Burbank; David Starr Jordan, president of Stanford University; and professor Charles Davenport of the University of Chicago.

Four years later, Davenport convinced Mrs. E. H. Harriman (the wife of the famous industrialist) to purchase a tract of land at Cold Spring Harbor, New York, where he established the Eugenics Record Office. According to Davenport, Mrs. Harriman's enthusiasm for the program was due to "the fact that she was brought up among well-bred race horses [which] helped her to appreciate the importance of a project to study heredity and good breeding in Man."[8] Davenport, who was director, and Harry H. Laughlin, the superintendent, were soon to become the dominant voices in the American eugenics movement.

After 1910, eugenics societies sprang up in cities all over the country. Among the most influential were the Galton Society of New York and the Eugenic Education Societies of Chicago; St. Louis; Madison, Wisconsin; Battle Creek, Michigan; and San Francisco.[9] In 1913, the Eugenics Association was established, and in 1922 the Eugenics Committee of the United States (later the American Eugenics Society) was formed.[10]

By World War I eugenics was a favorite topic not only in the schools and at political forums, but also at women's clubs, church meetings, and in popular magazines of the day. Eugenics dogma often bordered on near hysteria, as when the president of the University of Arizona warned that it is "an optimist indeed who can see in our trend toward race degeneracy . . . anything other than a plight in which the race must find its final destiny in trained imbecility."[11]

Some even began to call for a basic change in our form of government to accommodate a eugenics ideology. William McDougall, chairman of the psychology department of Harvard University, so feared that democracy would eventually result in the "lower breeds" outnumbering the "best stock" and overtaking the machinery of the state that he openly advocated a caste system for America, based on biological differences, in which political rights would depend on one's caste.[12]

Academics were so convinced of the wisdom and virtue of applied eugenics that many of them threw all scholarly caution to the wind. "We know enough about eugenics," said Charles R. Van Hise, president of the University of Wisconsin, "so that if the knowledge were applied, the defective classes would disappear within a generation."[13]

Most of the leading educators of the day agreed with Irving Fisher, the well-known Yale economist, that "eugenics is incomparably the greatest concern of the human race."[14] It's not surprising, then, that by 1928 more than three-fourths of all the colleges and universities in America were teaching eugenics.[15] Their teachers were men like Earnest A. Hooton of Harvard, who preached that "crime is the result of the impact of environment upon low-grade human organisms." "The solution to the crime problem," he told Harvard undergraduates, is the "extirpation of the physically, mentally and morally unfit or (if that seems too harsh) their complete segregation in a socially aseptic environment."[16]

The eugenics creed also found willing adherents within the media. It might interest today's subscribers to the prestigious left-liberal magazines *The Nation* and *The New Republic* that the founders of both publications were crusaders for eugenics reform. Edwin Laurence Godkin, founder of *The Nation,* believed that only those of superior biological stock should run the affairs of the country,[17] and Herbert David Croly of *The New Republic* was convinced that blacks "were a race possessed of moral and intellectual qualities inferior to those of the white man."[18]

Imagine, if you will, a future president of the United States quoted in *Good Housekeeping* magazine to the effect that "there are racial considerations too grave to be brushed aside for any sentimental reasons." According to President Coolidge, biological laws tell us that certain divergent peoples will not mix or blend. Coolidge concludes that the Nordics propagate themselves successfully, "while with other races, the outcome shows deterioration on both sides."[19]

Some of America's great heroes also succumbed to the eugenics fervor. Alexander Graham Bell was one of them. Speaking before the American Breeders Association in Washington in 1908, Bell remarked: "We have learned to apply the laws of heredity so as to modify and improve our breeds of domestic animals. Can the knowledge and experience so gained be available to man, so as to enable him to improve the species to which he himself belongs?" Bell believed that "students of genetics possess the knowledge . . . to improve the race" and that education of the public was necessary to gain acceptance for eugenics policies.[20]

Even the fledgling Boy Scout movement in America was affected by the eugenics fervor. David Starr Jordan, who served as vice-president of the Boy Scouts of America in those early days, believed that the Scout program could help rear the "eugenic new man."[21]

Many modern-day feminists will be chagrined to learn that Margaret Sanger, a leader in the fight for birth-control programs, was a true believer in the biological superiority and inferiority of different groups. In some of the strongest words to ever come out of the eugenics movement, Sanger remarked that, "It is a curious but neglected fact that the very types which in all kindness should be obliterated from the human stock, have been permitted to reproduce themselves and to perpetuate their group, succored by the policy of indiscriminate charity of warm hearts uncontrolled by cool heads." Sanger had her own ideas about how to rid society of the problem of human biological contamination and promote better breeding. She wrote, "There is only one reply to a request for a higher birth rate among the intelligent and that is to ask the government to first take the burden of the insane and feebleminded from your back. . . . Sterilization is the solution."[22]

Eugenicists looked on sterilization as a major tool in their efforts to weed out biologically inferior stock from the American population. Their relentless lobbying campaigns succeeded. Tens of thousands of American citizens were involuntarily sterilized under various laws enacted by the individual states after the turn of the century. Indiana passed the first sterilization law in 1907. The statute called for mandatory sterilization of confirmed criminals, idiots, imbeciles, and others in state institutions when approved by a board of experts. It was later fondly referred to by eugenicists as "the Indiana Idea."[23]

The newfound willingness to believe in the hereditary basis of dependency, delinquency, and crime was due, in part, to the public's frustration with the failed efforts at social reform. One reformatory superintendent noted that experts were being pressured by the public to explain the apparent inability of state institutions to rehabilitate. He observed, "The only way in which their criticism can be met is by producing data showing that a large majority of these failures were due to mental defect on the part of the inmates and not to faults in the system of training."[24] Once this line of thought was accepted, sterilization became the easiest and most logical solution to the problem.

In 1914, Harry H. Laughlin issued a report to the American Breeders Association in which he stated, "Society must look upon germ plasm as belonging to society and not solely to the individual who carries it." The most startling part of his report was the finding that 10 percent of the population of the United States were "socially inadequate biological varieties" who should be segregated from the federal population and sterilized.[25]

With demands for sterilization mounting, fifteen more states enacted laws between 1907 and World War I. The extent to which the sterilization mania was carried is reflected in a bill introduced in the Missouri legislature calling for sterilization of those "convicted of murder, rape, highway robbery, chicken stealing, bombing, or theft of automobiles."[26]

The constitutionality of these sterilization laws was tested and upheld in 1927, when the U.S. Supreme Court ruled in a Virginia case that sterilization fell within the police powers of the state. The esteemed jurist Oliver Wendell Holmes wrote:

> We have seen more than once that the public welfare may call upon the best citizens for their lives. It would be strange if it could not call upon those who already sap the strength of the state for these lesser sacrifices, often felt to be such by those concerned, in order to prevent our being swamped with incompetence. It is better for all the world, if instead of waiting for their imbecility, society can prevent those who are manifestly unfit from continuing their kind . . . three generations of imbeciles are enough.[27]

By 1931, thirty states had passed sterilization laws and tens of thousands of American citizens had been surgically "fixed."[28]

## Importing the Best Blood

The eugenicists' greatest triumph came after World War I in their successful campaign to enact an immigration law based on eugenics standards. The law, which was passed in 1924, and which remained on the books until 1965, had the effect of altering the entire ethnic and racial composition of the United States to satisfy the standards laid down by eugenics supporters.

Eugenicists, early on, saw the value of mounting an ambitious drive to restrict immigration along "biological" lines. Writing to Charles Davenport in 1912, Irving Fisher of Yale remarked: "Eugenics can never amount to anything practically until it has begun, as Galton wanted it, to be a popular movement with a certain amount of religious fervor in it and as . . . there is already a sentiment in favor of restricting immigration . . . this is a golden opportunity to get people in general to talk eugenics."[29]

The opportunity to mount an all-out public drive for restrictive immigration came after World War I. The new mood of nationalism and isolationism, renewed labor strife, fear of a Marxist takeover of the country in the wake of the Russian Revolution, and the hordes of immigrants streaming into Ellis Island all created an ideal atmosphere for eugenics legislation. Overnight, eugenicists began cranking out a new series of studies and reports to demonstrate the biological inferiority of certain immigrant groups.

Liberal sociologist Edward A. Ross, after a sixteen-month study, published a report entitled *The Old World in the New,* in which he claimed, among other things, that Mediterranean peoples were prone to sex and violence and were irrational by nature; the Slavs were a passive people imbued with ignorance and superstition, the men wife beaters and alcoholics; and the Jews were clannish, tricky and underhanded in business.[30] Another eugenicist, Madison Grant, added to the list the Hindus, "who have been for ages in contact with the highest civilizations, but have failed to benefit by such contact either physically, intellectually or morally" and the Negroes, who "are the Nordics' willing followers and who ask only to obey and to further the ideals and wishes of the master race."[31] The Nordics, on the other hand, were found to be a race of "great energy and industry, vigorous, imaginative and highly intelligent."[32]

The nerve center of the eugenics campaign for restrictive legislation was the House Committee on Immigration and Naturalization. If there were ever any question as to where the committee stood on the matter of immigration, it was quickly put to rest with the appointment of Harry H. Laughlin as the "Expert Eugenics Agent" of the committee, a post which allowed him to control the course of the proceedings.

Laughlin told the committee, "Making all logical allowances for environmental conditions . . . the recent immigration as a whole, presents a higher percentage of inborn socially inadequate qualities than do the older stocks."[33] Throughout the hearings, no one was called to refute Laughlin's findings.

The Secretary of Labor at the time, James J. Davis, expressed the sentiments of the Coolidge administration on the question of restrictive immigration, and, in so doing, set the tone for the ensuing debate: "America has always prided itself upon having for its basic stock the so-called Nordic race . . . we should ban from our shores all races which are not naturalizable under the law of the land and all individuals of all races who are physically, mentally, morally and spiritually undesirable and who constitute a menace to our civilization."[34]

The debate in the House of Representatives ended up being little more than a testimonial to the eugenics movement that had swept over the country. Congressman Robert Allen of West Virginia informed his colleagues, "The primary reason for the restriction of the alien stream is the necessity for purifying and keeping pure the blood of America."[35] J. Will Taylor of Tennessee warned, "America is slipping and sinking as Rome did and for identical causes. Rome had faith in the melting pot as we have. It scorned the iron certainties of heredity, as we do. It lost its instinct for race preservation, as we have lost ours."[36] Thomas V. Phillips of Pennsylvania said: "We know better than to import vicious or refractory animals but, on the contrary, through intelligent and careful selection from abroad, bend every effort to improve our homestock of domestic animals. . . . We must set up artificial means through legal machinery to hand pick our immigrants if we are to prevent rapid deterioration of our citizenship."[37]

Dozens of Representatives took to the floor to reaffirm their commitment to a "eugenic America." The opposition forces were small and disorganized, composed chiefly of Congressmen of Jewish, Italian, and other ethnic backgrounds. Their voices were all but drowned out amid the clamor to "purify the blood of the nation."

In the end, the new immigration legislation passed the House with only thirty-five Republicans and thirty-six Democrats voting against it. President Coolidge signed the legislation, which called for restrictions based on 2 percent of the foreign-born from each country according to the 1890 census. Since the number of Southern Europeans coming into the United States in 1924 was far greater than the number of Northern Europeans, this law had the effect of closing the door on these ethnic groups.

Emanuel Celler, a freshman in the 68th Congress, was one of the voices in opposition to the legislation. Jewish and representing a large Jewish and otherwise ethnic district in New York, Celler was appalled by the attitudes expressed during that debate. More than forty years later he was a prime architect in getting the legislation changed. In the interim, the American government operated under a eugenics standard in its overall immigration policy.

From its peak in 1924, the eugenics movement steadily declined, eventually collapsing five years later. The stock market crash of 1929 had a profound impact on eugenics philosophy. With America's financial elite jumping out of windows and middle-class professionals and academics standing in unemployment lines alongside Italian, Polish, and Jewish immigrants, it was no longer possible to maintain the myth that there was something biologically superior about certain kinds of people. The Depression served as the great equalizer as millions of Americans—Nordic and Italian, WASP and Jew—found themselves in the same circumstances of poverty and destitution. Biological distinction gave way to a shared sense of common plight for all those forced into the bread lines of the Depression.

Hitler's rise to power in Europe was the other major factor behind the decline in importance of the American eugenics movement. Ironically, in 1925, H. H. Newman, professor emeritus of zoology at the University of Chicago, penned the following:

> One needs only to recall the days of the Spanish Inquisition or of the Salem witchcraft prosecution to realize what fearful blunders human judgment is capable of, but it is unlikely that the world will ever see another great religious inquisition or that in applying to man the newly found laws of heredity there will ever be undertaken an equally deplorable eugenic inquisition.[38]

That same year, German officials were writing away to state governments in the United States for information on American sterilization laws. One of the leading advocates of eugenics in Germany at the time remarked, "What we racial hygienists promote is not at all new or unheard of. In a cultural nation of the first order, the United States of America, that which we strive toward was introduced long ago. It is all so clear and simple."[39]

While German eugenicists were busy reading reports about sterilization laws in the United States, the first printed edition of *Mein Kampf* was published in Germany. In it, Hitler proclaimed: "The mixing of higher and lower races is clearly against the intent of nature and involves the extinction of the higher Aryan race. . . . Wherever Aryan blood has mixed with that of lower peoples, the result has been the end of the bearers of culture."[40]

Hitler's Third Reich came to power in 1933. Almost immediately, the Minister of the Interior, Wilhelm Frick, announced to the world that "the fate of racial hygiene of the Third Reich and the German people will be indissolvably bound together."[41] On July 14, 1933, the Führer decreed the Heredity Health Law, a eugenics sterilization statute which was to be the first step in a mass eugenics program that would claim the lives of millions of people over the next twelve years. In response to Hitler's unfolding eugenics campaign, American eugenicists observed that Germany was "proceeding toward a policy that will accord with the best of thought of eugenicists in all civilized countries."[42]

Interestingly enough, throughout the 1930s the Genetics Society of America debated over and over again at its annual meetings the question of whether to formally condemn the eugenics policies of the Third Reich. There were never enough votes for such a condemnation.[43] In 1936, the University of Heidelberg awarded an honorary degree to Harry Laughlin for his great contribution to the field of eugenics.[44]

## A New "User Friendly" Eugenics

After World War II, many opponents of eugenics hoped that the eugenics movement had finally come to rest alongside the unmarked mass graves that scarred the European landscape. Their hopes were to be short-lived. In the 1970s, the world began to hear scattered reports of great scientific advances occurring in the new field of molecular biology. Some scientists at the time worried that genetic engineering might lead to a return to the kind of eugenics movement that had seduced America and Europe earlier in the century. Speaking at a National Academy of Science forum on recombinant DNA in 1977, Ethan Signer, a biologist at the Massachusetts Institute of Technology, warned his colleagues,

> This research is going to bring us one more step closer to genetic engineering of people. That's where they figure out how to have us produce children with ideal characteristics. . . . The last time around the ideal children had blond hair, blue eyes and Aryan genes.[45]

By the early 1990s, a torrent of breathtaking new discoveries and applications were being announced in the biotech field. Most people found themselves unprepared to assess the full social implications of the many new genetic breakthroughs that seemed to be challenging so many well-established customs and conventions. Today, scientists are developing the most powerful set of tools for manipulating the biological world ever conceived. The newfound power over the life force of the planet is, once again, raising the specter of a new eugenics movement. This is the troubling reality that so few policy makers, and even fewer biologists, are willing to acknowledge.

The new genetic engineering tools are, by definition, eugenics instruments. Whenever recombinant DNA, cell fusion, and other related techniques are used to "improve" the genetic blueprints of a microbe, plant, animal, or human being, a eugenics consideration is built into the process itself. In laboratories across the globe, molecular biologists are making daily choices about what genes to alter, insert, and delete from the hereditary code of various species. These are eugenics decisions. Every time a genetic change of this kind is made, the scientist, corporation, or state is implicitly, if not explicitly, making decisions about which are the good genes that should be inserted and preserved and which are the bad genes that should be altered or deleted. This is exactly what eugenics is all about. Genetic engineering is a technology designed to enhance the genetic inheritance of living things by manipulating their genetic code.

Some might take offense at the suggestion that the new genetic engineering technology is reintroducing eugenics into our lives. They prefer to equate eugenics with the Nazi experience of more than five decades ago. The new eugenics movement bears little resemblance to the reign of terror that culminated in the Holocaust. In place of the shrill eugenic cries for racial purity, the new commercial eugenics talks in pragmatic terms of increased economic efficiency, better performance standards, and improvement in the quality of life. The old eugenics was steeped in political idealogy and motivated by fear and hate. The new eugenics is being spurred by market forces and consumer desire.

Is it wrong, ask today's molecular biologists, to want healthier babies? The new eugenics is coming to us not as a sinister plot, but rather as a social and economic boon. Still, try as we will, there is simply no way to get around the fact that the fledgling commercial effort to redesign the genetic blueprints of life on Earth is bringing us to the threshold of a new eugenics century.

## The Ultimate Therapy

For the first time in history, the scientific tools are becoming available to manipulate the genetic instructions in human cells. The new gene splicing techniques will make it potentially possible to transform individuals and future generations into "works of art," continually editing their DNA codes for therapeutic and cosmetic ends. As mentioned in chapter one, genetic manipulation is of two kinds. In somatic therapy, intervention takes place only within somatic cells and the genetic changes do not transfer into the offspring. In germ line therapy, genetic changes are made in the sperm, egg, or embryonic cells, and are passed along to future generations. Germ line experiments have been successfully carried out on mammals for more than a decade and researchers expect the first human trials to be conducted within the next several years. While somatic gene therapy is likely to become commonplace within the next decade, germ line therapy is not likely to become a widespread practice for fifteen to twenty years. Similarly, monogenic changes (involving a single gene) will likely be standard practice within the next ten years, while polygenic gene manipulation (involving clusters of genes responsible for specific diseases, dispositions, and disorders) is likely to become a reality by the year 2025.

The first human somatic therapy was conducted in 1990 by Dr. French Anderson and a team of scientists from the National Institutes of Health (NIH) on a young girl suffering from ADA deficiency or the so-called "bubble boy" disease. Children suffering from the genetic illness, also known as Severe Compromised Immuno-Deficiency disease or SCID, lack the enzyme adenosine deaminase and, as a result, have seriously compromised immune systems. The disease became identified as the "bubble boy" disease when the media publicized the story of a young boy named David who was forced to live his life inside a plastic bubble to protect him from germs.[46]

In the NIH experiment, white blood cells were taken from the girl's body and the gene that codes for adenosine deaminase was introduced into the cells. The genetically modified cells were then reintroduced back into the child's body using a modified animal retrovirus as the vector or vehicle of transmission. The media heralded the therapy as a great leap forward for medical science. Behind the scenes, however, scientists were less enthusiastic. What the public was never told was that a drug therapy had been successfully used to treat ADA-deficient youngsters for more than half a decade before the Anderson experiments, and that since that time none of the children suffering from the disease had to live inside protective bubbles. In fact, the child the NIH team experimented on was already being administered the ADA drug with some degree of effectiveness, which raised the question of how Anderson and his colleagues could ever assess whether it was the gene experiment or the drug that was responsible for the child's improved condition.[47]

Most molecular biologists failed to point out this major flaw in the experiment, preferring to bask in the media glow surrounding the first human gene therapy. A few biologists, however, did make known their opinions that the experiment was driven more by personal ambition and commercial gain than by good science. Dr. Stuart Orkin, a professor of pediatric medicine at Harvard Medical School, said at the time, "A large number of scientists believe the experiment is not well founded scientifically." Orkin admitted that he was personally "quite surprised that there hasn't been more of an outcry against the experiment by scientists who are completely objective." Dr. Richard Mulligan, a pioneer in gene therapy and member of the board of the NIH Recombinant DNA Advisory Committee, added, "If I had a daughter, no way I'd let her near these guys if she had that defect."[48]

There are several ways to intervene, at the somatic level, to cure genetic diseases. In some instances, a patient's genes may fail to produce a needed protein. With the new cloning techniques, it may be possible to identify, isolate, and produce large quantities of the missing protein, which could then be reintroduced back into the patient's body. In other instances, genes might be introduced into a patient in order to trigger an immune response and incapacitate precancerous cells.

The current technology for gene intervention, however, suffers from a number of drawbacks, the most important being the random nature of the procedure. Although the public has been led to believe that the first human gene therapy experiments have been carried out with a high degree of medical precision, the reality is quite the opposite. As in other animals, insertion of modified genes into a patient's chromosomes is random. Researchers cannot predict where on a chromosome the modified gene might land, raising the possibility of inadvertently disrupting other cellular functions. Even if a modified gene makes it to the desired location, there is no guarantee that it will express itself once there.[49] Dr. Phillip Kitcher, professor of philosophy at the University of California at San Diego, zeroes in on the primitive state of the art when he says, "Shooting some DNA into a patient's body, in the hope that it will be taken up by the right cells and will constantly churn out protein, is the molecular equivalent of hitting the body with a mallet, in the hope that a 'good whack' might just do the trick."[50]

Despite years of favorable media reports on various gene therapy experiments and the high expectations voiced by the medical establishment and the biotech industry, the results have, thus far, been so disappointing that the NIH itself was recently forced to acknowledge the fact and issue a sober warning to scientists conducting the experiments to stop making promises that cannot be kept. In an extensive survey of all 106 clinical trials of experimental gene therapies conducted over the past five years involving more than 597 patients, a panel of experts convened by the NIH reported that "clinical efficacy has not been definitively demonstrated at this time in any gene therapy protocol, despite anecdotal claims of successful therapy."[51] Even Dr. Leroy B. Walters, a philosophy professor at Georgetown University and the chairperson of the NIH oversight committee that reviewed and approved all of the clinical trials, remarked in a moment of candor that he and the committee had not seen "any solid results yet" after years of experiments.[52] Still, many of the staunchest supporters of the new gene therapies remain convinced that the techniques will bear fruit as methodologies and procedures are honed and new knowledge of the workings of the genes becomes available to researchers and clinicians.

Far more controversial is the prospect of conducting human germ line therapy. Debate over genetic manipulation of human eggs, sperm, and embryonic cells has raged for more than fifteen years. In 1983, a cross-section of the nation's religious leaders and prominent scientists announced their opposition to such experiments, on eugenics grounds, and urged a worldwide ban. (The coalition was organized by The Foundation on Economic Trends.)

The possibility of programming genetic changes into the human germ line to direct the evolutionary development of future generations brings society to the precipice of a eugenics era whose consequences, both to the biology of our species and to civilization, are unpredictable and unknowable. Even so, a growing number of molecular biologists, medical practitioners, and pharmaceutical companies are anxious to take the gamble, convinced that manipulating our evolutionary destiny is humankind's next great frontier. Their arguments are couched in terms of personal health, individual choice, and collective responsibility for future generations.

Writing in *The Journal of Medicine and Philosophy,* Dr. Burke Zimmerman makes several points in defense of germ line therapy over somatic cell therapy. To begin with, he argues that if and when somatic therapy becomes successful, it will likely increase the number of survivors with defective genes in their germ lines—genes that will continue to accumulate and further "pollute" the genetic pool of the species, passing an increasing number of genetic problems onto succeeding generations. Secondly, although somatic therapy may be able to treat many disorders in which treatment lies in replacing populations of cells, it may never prove effective in addressing diseases involving solid tissues, organs, and functions dependent on structure—the brain, for example—and therefore, germ line therapy likely will be the only remedy, short of abortion, for such disorders.[53]

Zimmerman and other proponents of germ line therapy argue for a broadening of the healing profession's ethical mandate to include responsibility for the health of those not yet conceived. The interests of the patient, they say, should be extended to include the interests of "the entire genetic legacy that may result from intervention in the germ line."[54] Moreover, parents ought not to be denied their right as parents to make choices on how best to protect the health of their unborn children during pregnancy. To deny them the opportunity to take corrective action in the sex cells or in the embryonic cells would be a serious breach of medical responsibility. Proponents of germ line therapy ask why countless individuals should be subjected to possibly painful, intrusive, and potentially risky somatic therapy when the gene or genes responsible for their diseases could be more easily eliminated from the germ line, at less expense, and with less discomfort.[55]

Finally, the health costs to society need to be factored into the equation, say the advocates of germ line therapy. Although the costs of genetic intervention into the germ line to cure diseases will likely be high in the early years, the cost is likely to drop dramatically in the future as the methods and techniques become more refined. Alternatively, the lifetime cost of caring for generations of patients suffering from Parkinson's disease or severe Down syndrome, they say, is likely to be far greater than simple prevention in the form of genetic intervention at the germ line level.

## The Slippery Slope

Prenatal testing has already laid much of the philosophical groundwork for the acceptance of eventual germ line genetic intervention and a new commercial eugenics era. Hundreds of thousands of pregnant women routinely test their unborn children in the womb for a wide range of genetic diseases.[56] Amniocentesis, the oldest and most widely used method, dates back to the 1960s. The physician inserts a needle into the womb and extracts some of the amniotic fluid, which contains fetal cells. The cells are tested for more than one hundred and fifty disorders. A newer procedure, chorionic villus sampling, allows pregnant women to test for defects as early as the tenth week after conception, in contrast to amniocentesis which can be used only between the fourteenth and twentieth week of pregnancy. With the newer method, the physician conducts a biopsy of fetal membrane tissue for genetic analysis.[57]

Although the prenatal tests can detect many genetic disorders, as of now, less than 15 percent of those disorders can be treated. This means, for the vast number of seriously debilitating or fatal diseases that are testable in the womb, the only choices are elective abortion or bringing the baby to term. The pressure on a pregnant woman to either abort her unborn child or give birth to a baby suffering from Down syndrome, spina bifida, Turner's syndrome, Tay-Sachs disease, or sickle-cell anemia can be agonizing. Choosing to end the new life in the womb or give birth to a child who might experience intense suffering and premature death has created a spate of new ethical dilemmas for which there are no historical precedents and few easy answers. The decisions are made more difficult by the fact that there are fewer than a thousand medical geneticists in the United States—and far fewer in other countries—who can help counsel prospective parents who must make critical life and death choices for their offspring.[58]

Decisions are complicated by the fact that some of the genetic diseases show up in infancy and are 100 percent fatal, like Tay-Sachs disease,

Duchenne's muscular dystrophy, and Gaucher's disease, while others strike later in life, like Huntington's disease or polycystic kidney disease. Should, for example, a pregnant woman abort or bring to term her unborn child if she knows the child will develop Huntington's disease, but not until midlife?

Then, too, many of the genetic disorders being tested, like spina bifida and Down syndrome, manifest varying degrees of severity, ranging from slight to extreme debilitation. How do prospective parents factor in the unknown possibilities and make decisions about what the appropriate threshold might be for an acceptable quality of life for their child? A good case in point is Down syndrome. Children inheriting an extra chromosome 21 can end up being severely or only moderately retarded. A nurturing and stimulating environment can also have some effect on a Down syndrome child's future development.[59]

The confusion, doubts, and uncertainties over how genetic disorders will manifest themselves, as well as the role environment plays in development, can lead to uninformed choices for thousands of prospective parents. An unborn child, for example, might carry a defective gene and still never manifest the disease during his or her lifetime. In the 1970s, researchers discovered that some men carry an extra Y chromosome, leading scientists to speculate that it might precondition the affected male to be more aggressive and potentially antisocial. When serial killer Richard Speck was found to carry an extra Y chromosome, researchers began screening the prison population for the so-called "criminal" trait. Studies found a high percentage of inmates in mental hospitals with the extra Y chromosome. It is impossible to know how many pregnant mothers may have aborted a male child upon finding he harbored the extra chromosome. Subsequent studies, however, have revealed that a large number of men in the general population also have the extra Y chromosome and carry on normal lives, thus casting considerable doubt over the relevance of the "defect."[60]

As already mentioned, until very recently, the only options open to prospective parents faced with an unborn child with a genetic disorder have been elective abortion or bringing the child to term. Now, a new cutting-edge technology called pre-implantation genetics allows parents an alternative choice. The first live birth of a child resulting from pre-implantation genetics screening occurred in March of 1992 in London.

The child, named Chloe, had been genetically screened at the embryo stage for cystic fibrosis (CF) before being implanted into the mother's womb.

A medical team first removed several eggs from the mother's womb and fertilized them in vitro, with the husband's sperm. After the eighth cell division, researchers pierced each embryo with a hollow instrument and extracted a cell. Each cell was then copied several thousand times using a technology called polymerase chain reaction. Scientists screened each lot for the CF gene and discarded those embryos found to harbor the defect. Two embryos were subsequently implanted in the mother, one with two normal genes, and one with a normal gene and a CF gene. (Bear in mind that two CF genes are required to manifest the disease.) One of the embryos developed, and Chloe was born free of the CF disease. Researchers in several countries are now proceeding with pre-implantation genetic screening for other diseases, including sickle-cell anemia and Tay-Sachs disease.[61]

"Parental eugenics," in the form of pre-implantation screening of embryos for defects, is likely to increase significantly in the near future as more screening tests become available and embryo implantation procedures become cheaper and more reliable. Still, these first rather crude procedures are only the beginning of a wholesale change in the relationship between parents and their unborn children that is likely to redefine the very notion of parenthood in the Biotech Century.

Over the next ten years, molecular biologists say they will locate specific genes associated with several thousand genetic diseases. In the past, a parent's genetic history provided some clues to genetic inheritance, but there was still no way to know for sure whether specific genetic traits would be passed on. In the future, the guesswork will be increasingly eliminated, posing a moral dilemma for prospective parents. Parents will have at their disposal an increasingly accurate readout of their individual genetic makeups, and will be able to predict the statistical probability of a specific genetic disorder being passed on to their children as a result of their biological union.

To avoid the emotional anguish of such decisions, some young people are likely to opt for prevention and avoid marrying someone of the wrong "genotype" for fear of passing along serious genetic diseases to their offspring.[62] Already, part of the Orthodox Jewish community in the United States has established a nationwide program to screen all young Jewish men and women for Tay-Sachs disease. Every young Jew is encouraged to take the test. The results are made available in an easily accessible database to allow young eligible men and women to choose their prospective spouses with genotype in mind.

Some ethicists argue that such programs will become far more commonplace, placing a "genetic stigma" on young people. There's ample precedent for concern. Researchers report that when sickle-cell anemia was screened for in Greece, nearly 23 percent of the population was found to have the trait. Fearing stigmatization, many of the carriers concealed their test results, believing that public exposure would seriously jeopardize their marriage prospects.[63]

When researchers at Johns Hopkins Medical Center in Baltimore recently discovered a genetic alteration in one out of every six Jews of Eastern European ancestry that doubles their risk of getting colon cancer, many in the Jewish community began to express their concern that the Jewish population might be singled out and made the object of discrimination. The news of the "Jewish" cancer gene came on top of other discoveries linking breast and ovarian cancer, cystic fibrosis, Tay-Sachs, Gaucher's, and Canavan's disease to Jewish blood lines. Of course, scientists point out that other groups are likely to have just as many genetic links to specific diseases but that the Jewish population has received the most attention to date because "they constitute a well defined, easily identifiable and closely related community—exactly the kind that allows geneticists to start identifying disease-causing genes."[64] Still, the explanations of the researchers were not enough to calm an anxious Jewish community. Amy Rutkin, the director of American affairs for Hadassah, the nation's largest Jewish membership organization, reported that in the aftermath of the colon cancer discovery, the organization has been "receiving phone calls indicating a certain amount of fear and confusion." Rutkin said, "People are asking, is too much research focused on the Jewish community and are we at risk of stigmatization?"[65]

Health professionals also worry about genetic stigmatization and especially the prospect of selecting potential mates based on genotyping, but argue that it is still less onerous than elective abortion or sentencing a newborn to premature death or a life of chronic or debilitating illness. Not surprising, there is increasing talk of government-mandated genetic testing of couples seeking marriage licenses. Even without a government requirement, it's likely that a growing number of potential marriage partners will want their future partner screened before committing themselves to a lifelong relationship. For example, consider the hypothetical case of a man whose fiancée's family has a history of breast cancer and death in midlife. Before committing himself to marriage and a family, he might not think it unreasonable to ask her if she has been screened for the BRCA-1 gene (breast cancer gene). If there is a fair chance she will develop the disease in her early forties and die while her children are still young, he might consider that to be too great a burden to bear and choose not to marry her.

## Genetic Responsibility

In the coming decades, scientists will learn more about how genes function. They will become increasingly adept at turning genes "on" and "off." They will become more sophisticated in the techniques of recombining genes and altering genetic codes. At every step of the way, conscious decisions will have to be made as to which kinds of permanent changes in the biological codes of life are worth pursuing and which are not. A society and civilization steeped in "engineering" the gene pool of the planet cannot possibly hope to escape the kind of ongoing eugenics decisions that go hand in hand with each new advance in biotechnology. There will be enormous social pressure to conform with the underlying logic of genetic engineering, especially when it comes to its human applications.

Parents in the Biotech Century will be increasingly forced to decide whether to take their chances with the traditional genetic lottery and use their own unaltered eggs and sperm knowing their children may inherit some "undesirable" traits, or undergo corrective gene changes on their sperm, eggs, embryos, or fetus, or substitute egg or sperm from a donor through *in vitro* fertilization and surrogacy arrangements. If they choose to go with the traditional approach and let genetic fate determine their child's biological destiny, they could find themselves culpable if something goes dreadfully wrong in the developing fetus, something they could have avoided had they availed themselves of corrective genetic intervention at the sex cell or embryo stage.

Consider the following scenario. Two parents decide not to genetically "program" their fetus. The child is born with a deadly genetic disease and dies prematurely and needlessly. The genetic trait responsible for the disease could have been deleted from the fertilized egg by simple gene surgery. In the Biotech Century, parents' failure to correct genetic defects *in utero* may well be regarded as a heinous crime. Society could conclude that every parent has a responsibility to provide as safe and secure an environment as humanly possible for their unborn child. Not to do so might be considered a breach of parental duty for which the parents could be held morally, if not legally, liable. Mothers have already been held liable for having given birth to crack cocaine–addicted babies and babies with fetal alcohol syndrome. Prosecutors have argued that mothers passing on these painful addictions to their unborn children are culpable under existing child abuse statutes, and ought to be held liable for the effect of their lifestyle on their babies.

Even more ominously, "wrongful life" and "wrongful birth" lawsuits have begun to appear in the United States. More than three hundred such cases have made their way through the courts. In the case of "wrongful birth" lawsuits, parents of a seriously ill or disabled child sue their physician or hospital claiming the child should never have been born. The lawsuits charge negligence on the part of the health provider for not advising parents of a potential health problem with their unborn, and of not making available information on screening procedures which could have been performed and whose results could have been used to make an informed decision on whether or not to abort the fetus. In "wrongful life" lawsuits, the claim is brought on behalf of the child or by the child, claiming that he or she should never have been born. While the current spate of lawsuits are aimed at attending physicians, it's not unlikely that in the case of "wrongful life" lawsuits, children, in the future, might similarly charge their parents with negligence for not performing the appropriate screening tests or for ignoring the results of the screens and bringing the baby to term.

One of the most interesting of these suits was brought in 1975 by Paul and Shirley Berman against two New Jersey doctors. The Bermans' daughter, Sharon, was born with Down syndrome. The Bermans argued that the doctors were negligent in not advising them of the desirability of undergoing amniocentesis, despite the fact that Mrs. Berman was thirty-eight years old at the time of her pregnancy, and therefore at risk of having a Down syndrome child.

The Bermans contended that had they known that they were carrying a child with Down syndrome, they would have had an abortion. They filed a claim of "wrongful life," seeking compensation for the suffering the child would experience during her lifetime, as well as a claim for "wrongful birth" to compensate for their own "emotional anguish."

The New Jersey Supreme Court denied the Bermans' claim of "wrongful life," arguing that the claim was "metaphysical" in nature, and that the court could not be put in a position of judging "the difference in value between life in an impaired condition and the utter void of nonexistence." In his opinion, Justice Morris Pashman wrote, "Ultimately the infant's complaint is that she would be better off not to have been born. Man, who knows nothing of death or nothingness, cannot possibly know whether that is so."[66] (The New Jersey Supreme Court did award the Bermans "emotional damages," however, for their "wrongful birth" claim for the "mental and emotional anguish" caused by having a child with Down syndrome.) While six states have passed laws limiting or forbidding compensation for "wrongful life" claims, several courts have recognized a child's right to bring such cases.[67]

It is likely that as new screening technologies become more universally available, and genetic surgery at the embryonic and fetal stages becomes more widely acceptable, the issue of parental responsibility will be hotly debated, both in the courts and in the legislatures. The very fact that parents will increasingly be able to intervene to ensure the health of their child before birth is likely to raise the concomitant issue of the responsibilities and obligations to their unborn children. Why shouldn't parents be held responsible for taking proper care of their unborn child? For that matter, why shouldn't parents be held liable for neglecting their child's welfare in the womb in cases where they failed to or refused to screen for and correct genetic defects that could prove harmful to their offspring?

## Customizing Babies

Proponents of human genetic engineering argue that it would be cruel and irresponsible not to use this powerful new technology to eliminate serious "genetic disorders." The problem with this argument, says *The New York Times* in an editorial entitled "Whether to Make Perfect Humans," is that "there is no discernible line to be drawn between making inheritable repair of genetic defects and improving the species."[68] The *Times* rightly points out that once scientists are able to repair genetic defects, "it will become much harder to argue against additional genes that confer desired qualities, like better health, looks or brains."[69]

If diabetes, sickle-cell anemia, and cancer are to be prevented by altering the genetic makeup of individuals, why not proceed to other less serious "disorders": myopia, color blindness, dyslexia, obesity, left-handedness? Indeed, what is to preclude a society from deciding that a certain skin color is a disorder? In the end, why would we ever say no to any alteration of the genetic code that might enhance the well-being of our offspring? It is difficult to imagine parents rejecting genetic modifications that promised to improve, in some way, the opportunities for their progeny.

Prospective parents are already using genetic screening tests of their unborn children for purposes other than identifying debilitating diseases, suggesting that genetic intervention in the womb will likely be used in the future as much for whim and enhancement as for prevention or cure of illness. In 1955, researchers announced a successful procedure for determining the sex of a child by observing cells in the amniotic fluid. The test was first used in the 1960s for pregnant women with a history of hemophilia in their family. As males are the ones generally afflicted with the disease, pregnant women could elect to abort a male fetus. With the widespread use of amniocentesis in the 1970s, screening for sex became routine and doctors began reporting women selectively aborting normal male and female fetuses as a means of ensuring sibling gender balance in their families.[70]

Recent surveys have found increasing support for non-therapeutic "value preference" abortions. In one study, researchers reported that 11 percent of couples would abort a fetus that was predisposed to obesity.[71] The adoption of germ line therapy in the coming years is likely to shift emphasis from abortion to enhancement as increasing numbers of parents choose to "correct" cosmetic defects in the egg, sperm, or embryonic cells to ensure the birth of the best baby that medical science can produce.

We already have a strong inkling of how far parents might be willing to go to genetically enhance their children. The introduction of genetically engineered human growth hormone (hGH) has transformed the enhancement issue from an abstract intellectual concern to a hotly debated issue over public policy. In the 1980s both the Genentech and Eli Lilly companies were awarded patents to market a new genetically engineered growth hormone to the few thousand children suffering from dwarfism in the United States. The perceived market for the hormone was considered so small that both companies were awarded "orphan" drug status. This special designation allows the drug companies a number of compensations, including a monopoly over the sale of their product for seven years, as a reward for their willingness to invest so heavily in a drug designed to have such a limited market potential. By 1991, however, genetically engineered growth hormone had far eclipsed its original market expectations, becoming one of the best-selling pharmaceutical drugs in the country. Eli Lilly and Genentech now share a market of nearly $500 million in sales, making human growth hormone one of the greatest commercial success stories in the history of the pharmaceutical industry.[72]

Since there are so few children in the United States suffering from dwarfism, it is obvious that doctors across the country have been prescribing the hormone for normal children, who, while they may be shorter than their peers, do not suffer from a growth hormone deficit. Where parents have been reluctant to ask for the hormone, the children have begun to on their own. Many young men have found the hormone helpful in building and sustaining muscle development and have been purchasing it illegally on the black market. One survey carried out in a suburb of Chicago reported that more than 5 percent of suburban tenth-grade boys were illegally taking the drug.[73]

Mindful that tall people generally do better in life—command higher salaries, attract more desirable mates, and enjoy other similar perks—many parents are anxious to add an additional few inches onto their children and are willing to pay the exorbitant price for the weekly injections if it will give their children a leg up on the competition.

To ensure an ever-expanding market for the genetically engineered hormone, both Genentech and Eli Lilly have mounted an aggressive public relations and marketing campaign—with the help of local physicians—to redefine normal shortness as an "illness." With the encouragement and financial support of the two drug companies, a number of researchers and pediatricians are arguing that children in the bottom 3 percent of height in their age group might be defined as abnormal and in need of growth hormone to catch up with their peers. If their assumptions become the orthodoxy among pediatricians and family physicians, genetically engineered growth hormone could be used by up to ninety thousand children born each year in the United States, with a potential market of eight to ten billion dollars in sales.[74]

Surprisingly, the NIH has weighed in on the side of the drug companies and is currently conducting a twelve-year research project—partially funded by Eli Lilly—to assess the effect of genetically engineered human growth hormone on short-statured children who do not suffer from growth hormone deficiency but were merely born into families with shortness in their family trees.[75] It should be pointed out that, by law, the NIH is prevented from experimenting on healthy children and exposing them to unnecessary health risks. The Foundation on Economic Trends petitioned the agency in a series of formal legal challenges designed to ferret out the Institutes' rationale for what appeared to be purely cosmetic or enhancement experiments on healthy children. Determined to find a legal justification for these precedent-setting non-therapeutic experiments, NIH argued, in the words of NIH spokesperson Micheala Richardson, "These kids are not normal. They are short in a society that looks at that unfavorably."[76]

Even *The Journal of the American Medical Association* was forced to concede the obvious. In an editorial, the *Journal* said, "We do not usually call prejudice-induced conditions, which confer cultural disadvantages but have no intrinsic negative health effects, diseases."[77]

Redefining short-statured children—who are not suffering from growth hormone deficiency—as abnormal simply because society looks unfavorably on short people punishes the victim of prejudice for not measuring up to the norms imposed by the majority. That the NIH could be party to such a flagrant and egregious violation of ethical standards speaks forcefully to the new eugenics wind blowing over the land—a eugenics motivated by the push of the marketplace and the pull of consumer desire for better, more perfect children.

Dr. John D. Lantos, of the Center for Clinical Medical Ethics at the University of Chicago, summed up the importance of the hGH marketing campaign in furthering a new commercially driven eugenics era. He observes,

> Until growth hormone came along, no one called normal shortness a disease. It's become a disease only because a manipulation [hGH] has become available and because doctors and insurance companies, in order to rationalize their actions, have had to perceive it as one. What we are seeing is two things—the commodization of drugs that are well-being enhancers and the creeping redefinition of what it means to be healthy.[78]

The robust market for hGH as an enhancement therapy is indicative of the vast commercial potential of genetic therapies for purposes that will likely transcend strictly medical uses. According to a 1992 Harris poll, 43 percent of Americans "would approve using gene therapy to improve babies' physical characteristics."[79] With Americans already spending billions of dollars on cosmetic surgery to improve their looks and psychotropic drugs to alter their mood and behavior, the use of genetic therapies to enhance their unborn children seems a likely prospect.

Many advocates of germ line intervention are already arguing for enhancement therapy. They contend that the current debate over corrective measures to address serious illnesses is too limited and urge a more expansive discussion to include the advantage of enhancement therapy as well. As to the oft-heard criticism that genetic enhancement will favor children of the rich at the expense of children of the poor—as the rich will be the only ones capable of paying for genetic enhancement of their offspring—proponents argue that the children of well-off parents have always enjoyed the advantages that wealth and inheritance confer. Is it such a leap, they ask, to want to pass along genetic gifts to their children along with material riches? Advocates ask us to consider the positive side of germ line enhancement, even if it gives an advantage to the children of those who can afford the technology. "What about . . . increasing the number of talented people. Wouldn't society be better off in the long run?" asks Dr. Burke Zimmerman.[80]

*The Economist* suggested in an editorial that society should move beyond old-fashioned hand-wringing moralism on the subject and openly embrace the new commercial eugenics opportunities that will soon become available in the marketplace. The editors asked,

> What of genes that might make a good body better, rather than make a bad one good? Should people be able to retrofit themselves with extra

> neuro-transmitters, to enhance various mental powers? Or to change the color of their skin? Or to help them run faster, or lift heavier weights?[81]

*The Economist* editorial board made clear that its own biases lay firmly with the marketplace. The new commercial eugenics, its editors argue, is about ensuring greater consumer freedom so that individuals can make of themselves and their heirs whatever they choose. The editorial concluded with a ringing endorsement of the new eugenics.

> The proper goal is to allow people as much choice as possible about what they do. To this end, making genes instruments of such freedom, rather than limits upon it, is a great step forward.[82]

Dr. Robert Sinsheimer, a long-standing leader and driving force in the field of molecular biology, laid out his eugenics vision of the new man and woman of the Biotech Century:

> The old dreams of the cultural perfection of man were always sharply constrained by his inherited imperfections and limitations. . . . To foster his better traits and to curb his worse by cultural means alone has always been, while clearly not impossible, in many instances most difficult. . . . We now glimpse another route—the chance to ease the internal strains and heal the internal flaws directly, to carry on and consciously perfect far beyond our present vision this remarkable product of two billion years of evolution. . . . The old eugenics would have required a continual selection for breeding of the fit, and a culling of the unfit. . . . The horizons of the new eugenics are in principle boundless—for we should have the potential to create new genes and new qualities yet undreamed. . . . Indeed, this concept marks a turning point in the whole evolution of life. For the first time in all time, a living creature understands its origin and can undertake to design its future. Even in the ancient myths man was constrained by essence. He could not rise above his nature to chart his destiny. Today we can envision that chance—and its dark companion of awesome choice and responsibility.[83]

## Perfecting the Code

While the notion of consumer choice would appear benign, the very idea of eliminating so-called genetic defects raises the troubling question of what is meant by the term "defective." Ethicist Daniel Callahan of the Hastings Center penetrates to the core of the problem when he observes, "Behind the human horror at genetic defectiveness lurks . . . an image of the perfect human being. The very language of 'defect,' 'abnormality,' 'disease,' and 'risk' presupposes such an image, a kind of proto-type of perfection."[84]

The all-consuming preoccupation with "defects" or "errors" among medical researchers and molecular biologists puts them very much at odds with most evolutionary biologists. When evolutionary biologists talk of "mutations," they have in mind the idea of "different 'readings' or 'versions' " of a relatively stable archetype.[85] James Watson and Francis Crick's discovery of the DNA double helix in the 1950s, however, brought with it a new set of metaphors and a new language for describing biological processes which changed the way molecular biologists perceive genetic mutations. The primary building block of life was described as a code, a set of instructions, a program, to be unraveled and read. The early molecular biologists, many of whom had been trained first as physicists, were enamored with what they regarded as the universal explanatory power of the information sciences. Norbert Wiener's cybernetic model and modern communications and information theory provided a compelling new linguistic paradigm for redefining how we talk about both physical and biological phenomena. (The new language of biology will be described in greater detail in Chapters Six and Seven.) It is within the context of this new language that molecular biologists first began to talk of genetic variation as "errors" in the code rather than "mutations." The shift from the notion of genetic mutations in nature to genetic errors in codes represents a sea change in the way biologists approach their discipline, with profound implications for how we structure both our relationship to the natural world and our own human nature in the coming Biotech Century.[86]

The significance of this shift in language became apparent to this author more than a decade ago during a debate on the future of biotechnology with the late Dr. Bernard Davis who was, at that time, chairman of the Department of Microbiology and Molecular Genetics at Harvard University. The debate took place at the annual conference of the American Association for the Advancement of Science. At one point during the discussion, the subject turned to the issue of human germ line therapy. I asked Dr. Davis if he supported the idea of using germ line therapy to eliminate the nearly four thousand or so monogenic diseases from the gene pool of the human race. His answer was an unqualified yes.

In his enthusiasm to correct the genetic programs, Davis had lost touch with the most elemental assumptions of evolutionary biology. What he perceived as errors to be fixed, more traditional biologists would view as variations on a theme—a rich reservoir of genetic diversity that is essential to maintaining the viability of a species against ever-changing environments and novel external challenges. We have learned, long ago, that recessive traits and mutations are essential players in the evolutionary schema. They are not mistakes, but rather options, some of which become opportunities. Eliminating so-called "bad" genes risks depleting the genetic pool and limiting future evolutionary options. Recessive gene traits are far too complex and mercurial to condemn as simple errors in the code. We are, in fact, just beginning to learn of the many subtle and varied roles recessive genes play, some of which have been critically important in insuring the survival of different ethnic and racial groups. For example, the sickle-cell recessive trait protects against malaria. The cystic fibrosis recessive gene may play a role in protecting against cholera. To think of recessive traits and single gene disorders, then, as merely errors in the code, in need of reprogramming, is to lose sight of how things really work in the biological kingdom.

Treating genetic disorders by eliminating them at the germ line level, in the sex cells, is far different from treating genetic disorders by way of somatic gene surgery after birth. In the former instance, the genetic deletions will result, in the long run, in a dangerous narrowing of the human gene pool upon which future generations rely for making evolutionary adaptations to changing environments. Somatic gene surgery, on the other hand, if it proves to be a safe, therapeutic way to treat serious diseases that cannot be effectively treated by more conventional approaches, would appear to have potential value.

The idea of engineering the human species—by making changes at the germ line level—is not too dissimilar from the idea of engineering a piece of machinery. An engineer is constantly in search of new ways to improve the performance of a machine. As soon as one set of defects is eliminated, the engineer immediately turns his attention to the next set of defects, always with the idea in mind of creating a more efficient machine. The very idea of setting arbitrary limits to how much "improvement" is acceptable is alien to the entire engineering conception.

The new language of the information sciences has transformed many molecular biologists from scientists to engineers, although they are, no doubt, little aware of the metamorphosis. When molecular biologists speak of mutations and genetic diseases as errors in the code, the implicit, if not explicit, assumption is that they should never have existed in the first place, that they are "bugs," or mistakes that need to be deprogrammed or corrected. The molecular biologist, in turn, becomes the computing engineer, the writer of codes, continually eliminating errors and reprogramming instructions to upgrade both the program and the performance. This is a dubious and dangerous role when we stop to consider that every human being brings with him or her a number of lethal recessive genes. Do we then come to see ourselves as miswired from the get-go, riddled with errors in our code? If that be the case, against what ideal norm of perfection are we to be measured? If every human being is made up of varying degrees of error, then we search in vain for the norm, the ideal. What makes the new language of molecular biology so subtly chilling is that it risks creating an unattainable new archetype, a flawless, errorless, perfect being to which to aspire—a new man and woman, like us, but without the warts and wrinkles, vulnerabilities and frailties, that have defined our essence from the very beginning of our existence.

No wonder so many in the disability rights community are becoming increasing frightened of the new biology. They wonder if, in the Biotech Century, people like themselves will be seen as errors in the code, mistakes to be eliminated, lives to be prevented from coming into being. Then again, how tolerant are the rest of us likely to be when we come to see others around us as defective, as mistakes and errors in the code?

The question, then, is whether or not humanity should begin the process of engineering future generations of human beings by technological design in the laboratory. What are the potential consequences of embarking on a course whose final goal is the "perfection" of the human species?

# Five The Sociology of the Gene

The extraordinary advances in genetic engineering are being accompanied by a resurgent eugenic sociology. The new sociology is helping create the cultural climate for the emergence of a eugenic society. Sociobiology has descended from the academic corridors of Harvard University and other elite educational institutions to become the social arbiter of the Genetic Era. In the age-old battle of nature vs. nurture, the sociobiologists side with the former school of thought, arguing that human behavior is more closely associated with one's biological inheritance than with one's environment. While the sociobiologists acknowledge that environment plays a role in individual and group development, they are far more impressed with the role that the genes play in determining one's sociality. The shift from nurture to nature is the fifth strand of the new operational matrix of the Biotech Century.

As knowledge about genes increases, the bioengineers will inevitably gain new insights into the functioning of more complex characteristics, such as those associated with behavior and thought. Researchers are already linking an increasing number of mental diseases to genetic disorders. Some scientists are even beginning to suggest that various forms of antisocial behavior, such as shyness, misanthropy, and criminality, may be examples of malfunctioning genes. The Minnesota Center for Twin and Adoption Research has found that heredity plays a determining role in a number of common personality traits. The center has even published studies estimating the extent to which heredity determines personality: tendency to worry, 55 percent; creativity, 55 percent; conformity, 60 percent; aggressiveness, 48 percent; extroversion, 61 percent.[1] Many sociobiologists go even further, contending that virtually all human activity is, in some way, determined by our genetic makeup, and that if we wish to change our situation, we must first change our genes.

## Nature over Nurture

Studies purporting to find the genetic basis for moods, behavior, and personality are proliferating in the scientific literature. In 1997, scientists at the Institute of Child Health in London reported that they had located what they believed to be a cluster of genes on the X chromosome that predisposes girls to better "social skills" than boys. The researchers found significant differences in social skills among two groups of girls who suffered from Turner's syndrome. The girls who inherited the paternal X chromosome were far more socially expressive, had an easier time making friends, got along better with family and teachers, and were found to be generally happier and better adjusted than those who inherited the X chromosome from their mother. The scientists concluded that when the X chromosome is inherited from the mother, it is apparently inactive, whereas when it is passed along from the father, it is active and fosters higher levels of social interaction.

The researchers asked the parents of eighty-eight girls with Turner's syndrome to evaluate their daughters for various behaviors. For example, were the girls "lacking in awareness of other people's feelings"? Did they "unknowingly offend people with their behavior"? Were they "very demanding of people's time"? Did they have "difficulty following commands unless they were carefully worded"? Then, using genetic analysis, they found the girls who inherited the X chromosome from their mothers had greater difficulty in expressing social skills than those girls who inherited the X chromosome from their fathers. (Normally, girls inherit two X chromosomes, one contributed by each parent, while boys inherit one X and one Y chromosome. Turner's syndrome girls inherit only a single X chromosome.)

When researchers studied "normal" boys and girls, they found that the boys, all of whom get their X chromosomes from their mothers, score low on social skills "just as the Turner's syndrome girls whose single X gene also passed down from their mothers."[2]

David H. Skuse, a research psychiatrist at the Institute of Child Development of University College in London and a member of the research team, said that the cluster of genes passed along from the father's X chromosome "may help a person infer what another person is feeling or thinking—in short, female intuition."[3] Skuse and his colleagues speculate that better social skills, including the ability to empathize and bond with others, may have conferred an evolutionary advantage on women, allowing them to compensate for their lack of physical brawn, while men's less sensitive psyches may have been an advantage in fighting and killing other men and animals.[4]

In another study published in 1996, researchers reported finding a genetic basis for "novelty seeking," "thrill seeking," and "excitability." Two separate research teams, led by Richard P. Ebstein of the Herzog Memorial Hospital in Jerusalem, Israel, and Jonathan Benjamin of the National Institute of Mental Health's Laboratory of Clinical Science, associated differences in novelty seeking and thrill seeking with lower levels of dopaminergic activity. Dopamine plays a critical role in stimulating euphoria. Researchers found that high levels of novelty-seeking behavior is "strongly associated with high plasma prolactin levels, which reflects low dopaminergic activity."[5]

Researchers at the National Institutes of Health (NIH) say they have located genes that predispose people to experience "high anxiety." Individuals who inherit one form of a gene on chromosome 17 are more likely to worry, says Dennis Murphy of the NIH in a study reported in *New Scientist* magazine. The gene linked to anxiety influences the production of a protein known as the serotonin transporter. This particular protein controls the level of serotonin in the brain, a chemical that affects mood. NIH scientists, working with a research team from the University of Würzburg in Germany, located a stretch of DNA near the transporter gene that is responsible for stimulating the gene into action. The promoter, a sequence of DNA which regulates the gene, can be inherited in two forms, one of which is twice as effective in stimulating the gene to produce the protein. A study of hundreds of volunteers found that individuals with the "sluggish" version of the gene promoter scored higher in psychological tests for neurotic behavior, including worrying, pessimism, and fear, than people with the more active version of the promoter. The researchers caution, however, that the genetic predisposition, while significant, probably accounts for less than 4 percent of the variation in neurotic behavior among people.[6]

Scientists are finding more and more links between behavior, personality, and genetic makeup. Dr. John J. Ratey, a psychiatrist at the Medfield State Hospital in Massachusetts, says that many people suffer from genetic-based "shadow disorders," mild forms of recognized neuropsychiatric disorders like attention deficit disorder, obsessive-compulsive disorder, depression, or mania. Some people unknowingly suffer from a subclinical form of autism, says Ratey. Autism is a neurological disorder that is characterized by individuals withdrawing into their own inner world and being unable to communicate and make emotional contact with others. Ratey found that many parents of autistic children manifest sub-threshold characteristics of the disease, even though they are never diagnosed with autism. Often, individuals suffering from subclinical autism exhibit odd behavior and are seen as "nerdy" or "weird" by others. They are inept at grasping social cues, and lack even the most rudimentary social graces. Others see them as incorrigible. Ratey believes that more significant genetic testing is likely to divulge a wide range of subtle genetic variations that help condition mood and behavior and shape individual personalities.[7]

It seems that every week or so a new study is published showing a likely connection between genotype and behavior. In 1993, the journal *Science* published a paper by five scientists at the National Cancer Institute reporting "a linkage between DNA markers on the X chromosome and male sexual orientation." Dr. Dean Hamer and his research team tracked the family history of seventy-six homosexual men and found that 13.5 percent of the brothers of the gay men were also homosexual—six times the probability for homosexuality among males. The study also found many more homosexual uncles and cousins on the maternal side of each family. The research team concluded that there is a 95.5 percent certainty that the predisposition for homosexuality exists in a gene, or several genes, near the end of the long arm of the X chromosome inherited from the mother.[8]

Earlier, Dr. Simon LeVay of the Salk Institute of Biological Studies in La Jolla, California, had claimed that differences between homosexual and heterosexual men were located at the base of the brain. Both reports raised the issue of whether homosexuality is a biological predisposition, social preference, an illness, a simple deviation from the norm, or a normal variation of sexuality.[9]

Another controversy arose in 1996 when scientists reported that they had located a genetic marker for alcoholism in mice. Dr. John Crabbe, a behavioral geneticist at the Veterans Administration's Medical Center in Portland, Oregon, and his colleagues reported that mice lacking the 5-HT1b receptor gene drank twice as much alcohol as normal mice when given a choice of drinking water spiked with ethanol or plain water. The mutant mice were also found to be more tolerant of alcohol and more aggressive. The 5-HT1b receptor binds to serotonin, a chemical used by the brain cells to communicate with each other. Crabbe was quick to point out that "alcoholism is not a single gene disease. About 50 percent is genetic and about 50 percent is environmental. And of the genetic part, there are many genes involved." Still, when the study was reported, the nuances and complexities of the genotype-phenotype relationship were reduced to the headline "A Gene for Boozy Mice."[10]

The "genetic factor" in family relationships is also the focus of increasing attention among scientists, psychologists, and family counselors. Researchers in behavioral genetics are exploring "genetic influences" within families with an eye toward understanding how those influences condition the relationship among family members. Writing in *The Journal of Marriage and the Family,* Dr. David Reiss, professor of psychiatry at George Washington University in Washington, D.C., argues that given the "fact" that "there is now considerable evidence of the hereditary nature of a broad range of cognitive and personality factors, as well as psychopathology," it is useful to begin establishing models for how "genetic influences" affect parent-child interactions, and interactions with siblings.[11] For example, Reiss contends that "self-perceived social confidence is heritable" and that if this genetic disposition exists for both parents and their progeny, the mutually reinforcing genetics could be a positive force "in influencing their children's peer relationships."[12] On the other hand, if a parent and child both have genetic predispositions for high levels of anxiety and stress, the constant interaction between the two is only likely to reinforce and exacerbate the level of stress each family member experiences. Reiss uses the example of a parent's "inability to anticipate and prepare a child for school circumstances or the parent's own anxiety about separating from the child." According to Reiss, the "same genes in the child lead to behaviors that their teachers rate as symptomatic."[13] The very idea of rethinking family relationships and child development by way of an interactive genetic model is radically different from the way family dynamics have been conceived of in the past. The genetic model represents a shift in what is meant by the very term family relationships.

## Genetically Correct Politics

The accumulating body of studies on the genetic links to personality and behavior is having an effect on public discourse. It is important to remember that from the end of World War II through the 1980s social scientists argued that it is only by instituting changes in the environment that social evils could be addressed. For more than forty years, the orthodox political wisdom favored nurture over nature. Now, plagued by deepening social crises, the industrial nations seem no longer able or willing to make significant changes by the traditional path of institutional and environmental reform. The sociobiologists and others of their persuasion contend that attempting to overhaul the economic and social system is at best palliative, and, at worst, an exercise in futility. The key to most social and economic behavior, they contend, is to be found at the genetic level. To change society, they claim, we must first be willing to change the genes, for, while the environment is a factor, the genes are ultimately the agents most responsible for shaping individual and group behavior.

The late Sir Julian Huxley expressed the convictions of a growing number of biologists and social scientists when he wrote:

> It is clear that for any major advance in national and international efficiency we can not depend on haphazard tinkering with social and political symptoms or ad hoc patching up of the world's political machinery, or even on improving education, but must rely increasingly on raising the genetic level of man's intellectual and practical abilities.[14]

Like Huxley and the American eugenicists in the early decades of the twentieth century, many scientists are becoming increasingly convinced that the problems lie not with the institutions humanity fashions, but with the way humanity itself is fashioned.

The radical shift from nurture to nature is attributable, in part, to the intense interest generated by the Human Genome Project and the many hyperbolic statements in the media by its most prominent advocates. James Watson, who served as the first director of the government-funded effort to decipher the human genome, summed up the enthusiasm of many of his colleagues involved in the multibillion-dollar government program. In an interview with *Time* magazine, Watson boldly asserted that "we used to think our fate was in our stars. Now we know, in large measure, our fate is in our genes." Others in the molecular biology community have been equally effusive. Biologist Walter Gilbert calls the Human Genome Project "The Holy Grail of Genetics," while biologist Norton Zinder refers to the human genome sequence as the "Rosetta Stone." Biologist Robert Sinsheimer goes even further, saying that the sequence "defines a human."[15]

Much of the rhetoric is no doubt politically motivated, designed to keep public attention focused on the great potential benefits that are likely to flow from the Human Genome Project. Assuring continued congressional funding for the project is likely never far from the thoughts of its champions in the fields of molecular biology and business who have much to gain financially from the genetic data being collected at the taxpayers' expense.

If "we are our genes," as an increasing number of molecular biologists argue—although always with the faint caveat that environment plays at least some small role in development—then the three-billion-dollar-plus investment in the Human Genome Project seems a small price to pay for vital information that both individuals and society as a whole can use to enhance our personal and collective well-being. The federal government's now defunct Office of Technology Assessment, in its evaluation of the merits of mapping and sequencing the human genome, expressed the sentiments of many both in government and the private sector when it concluded that "one of the strongest arguments for supporting human genome projects is that they will provide knowledge about the determinants of the human condition," including those diseases "that are at the root of many current societal problems."[16]

The very idea that the genes "are at the root of many current societal problems" would have been unthinkable and summarily dismissed by academicians and policymakers just a generation ago. When Arthur R. Jensen, a professor of education and psychology at the University of California at Berkeley, attempted to raise the genetic arguments in an article written in *The Harvard Educational Review* in 1969 entitled, "How Much Can We Boost IQ and Scholastic Achievement?," the American intellectual community responded with a vitriolic attack, warning that such thoughts were likely to bring us back to the eugenics fervor that spread throughout the country and around the world earlier in the century. Jensen responded to his critics by charging that reasonable debate was being "stifled [by the] zeitgeist of environmentalist egalitarianism."[17] Two years later, Nobel Laureate William Shockley, the inventor of the transistor, made equally audacious eugenics claims at the American Psychological Association's annual convention. Shockley told reporters at the conference that "diagnosis will . . . confirm that our nobly intended welfare programs are promoting dysgenics—retrogressive evolution through the disproportionate reproduction of the genetically disadvantaged."[18] When he suggested to the convention delegates that the government should encourage voluntary sterilization of persons of low intelligence, by way of financial incentives—the payment being proportional to the number of points below 100 that the individual scored on the IQ test—the intellectual establishment circled their wagons again, this time accusing Shockley of fanning the flames of racial hatred and bigotry.[19]

Today, what was regarded as heresy between the end of World War II and the late 1980s has gained increasing intellectual currency and is fast becoming orthodoxy, at least in the medical field. Although sterilization is still looked on askance, physicians and health professionals are increasingly counseling prospective mothers to abort fetuses who test positive for Down syndrome and other genetic disorders that might result in the birth of a child with an "unacceptably low" IQ.

Writing in the journal *Science,* editor Daniel Koshland agreed with the Office of Technology Assessment that a number of genetic diseases "are at the root of many social problems."[20] Koshland points to the growing ranks of the homeless, many of whom suffer from mental disease, as an example of the need to tackle social problems at their genetic roots, through preventive measures.[21] Koshland's genetic argument for homelessness left liberal social reformers aghast but raised few eyebrows within the scientific community, where many had already been won over to the genetic causation camp. The idea that homelessness might also have some relationship to the question of educational opportunity, income distribution policies of the marketplace, the marginalization of the workforce resulting from corporate downsizing and the introduction of new labor-saving technologies, and the increasing disenfranchisement of the poor, were curiously absent from Koshland's remarks, leaving the clear impression that our last best hope resides with the molecular biologists and their efforts to decipher the "Holy Grail" of biology—the human genome.

A few lone voices in the biology community continue to caution their colleagues that they are playing fast and loose with genetics and, in the process, providing grist for a new and potentially dangerous political agenda. Dr. Ruth Hubbard, professor emeritus of biology at Harvard University, warned in 1993 that the precipitous shift from nurture to nature is

> due in part to a conservative backlash against the gains of the civil rights and women's rights movements. These and similar movements have emphasized the importance of environment in shaping who we are, insisting that women, African Americans, and other kinds of people have an inferior status in American society because of prejudices against them, not because of natural inferiority. Conservatives are quick to hail scientific discoveries that seem to show innate differences which they can use to explain the current social order.[22]

If the social reformers of the 1950s, '60s, '70s, and '80s had given short shrift to the genetic basis of human development in their zeal to right the wrongs of society, the new genetic reformers seem poised at the opposite extreme, attributing far too much of human and social behavior to the genes.

The growing nature/nurture debate has polarized much of the academic community in recent years. However, biologists working in the new field of developmental genetics may provide some much-needed middle ground for understanding the many subtle relationships that exist between genotype and phenotype and between environmental triggers and genetic expression. Developmental geneticists would disagree with the widely held belief among biologists like Dr. Alexander Rich of the Massachusetts Institute of Technology and Dr. Sung Hou Kim of the University of California at Berkeley, who claim, "The instructions for the assembly and organization of a living system are embodied in the DNA molecules contained within the living cell."[23]

Biologists in the new field acknowledge that genes encode important information for the development of an organism, but they argue that they do not, in themselves, determine or control that development. Cell biologist Dr. Stuart Newman of the New York Medical College points out that living beings are "dynamical systems" and "are sensitive to inputs from their environments and, unlike machines, for example, can exhibit very different behavior and take on different forms under slightly different environmental conditions."[24] Newman says it's more appropriate to think of DNA as "a list of ingredients, not a recipe for their interactions."[25] He cites the example of a developing embryo in the womb:

> The "environment" of the genome includes not only externally controllable factors like temperature and nutrition, but also numerous maternally-provided proteins present in the egg cell at the time of fertilization. These proteins influence gene activity, and by virtue of variations in their amounts and spatial distribution in the egg can cause embryos even of genetically identical twins to develop in uniquely different ways.[26]

Newman and others in the field of developmental genetics argue that genes don't generate organisms. Rather, the very existence of genes already presupposes the existence of the organism in which they're embedded, and it is the organism itself that interprets, translates, and makes use of the genes in the course of its development. This is a far different approach to the working of genes than the reductionist argument of biologists like Richard Dawkins who contend that an organism is little more than the orchestrated program or field created by the genes.

The idea of genes as "master molecules" or "causal agents" is giving way to a more sophisticated understanding of genes as integral components of more complex networks that make up both an organism and its external environment. An article in *Scientific American* sums up much of the current thinking of biologists on the cutting edge of developmental genetics research.

> The mystery of how developing organisms choreograph the activity of their genes so that cells form and function at the right place and at the right time is now being solved. Hundreds of experiments have shown that organisms control much of their genes, most of the time, by regulating transcription.[27]

Despite the fact that new experimental research is undermining arguments and assumptions based on simple genetic reductionism, the idea of the "master molecule" that controls our biological destiny has proven so useful in advancing the interests of the molecular biologists and the many commercial firms that make up the biotech industry that it continues to gather momentum, both in the media and in public discourse, as an explanatory tool for understanding personality development, adolescent behavior, ethnic and racial differences, collective psychology, and even the working of culture, commerce, and politics.

Dr. Jonathan Beckwith, a professor of microbiology and genetics at Harvard University and one of the early pioneers in the field of molecular biology, argues that a more balanced presentation of the relationship between genetics and environment needs to be made in the public arena, lest we risk the new science becoming the handmaiden for a eugenics-based politics. Beckwith points out that many diseases such as cancer and depression are the result of the subtle and not so subtle interactions of genetic predispositions and environmental triggers, and to ignore the relationships and concentrate only on the gene is tantamount to abandoning any idea of moderating or reforming the environment as a remedial strategy. The political implications are significant.

> The focus on genetics alone as explanatory of disease and social problems tends to direct society's attention away from other means of dealing with such problems. . . . Genetic explanations for intelligence, sex role differences, or aggression lead to an absolving of society of any responsibility for its inequities, thus providing support for those who have an interest in maintaining these inequities.[28]

The troubling social and economic consequences of shifting to genetic causation as an all-encompassing explanatory model were dramatically illustrated several years ago in the firestorm of controversy that surrounded a federal government decision to co-sponsor a national forum on the genetic links to criminal behavior. The theme of the conference was "Genetic Factors in Crime" and was funded by the NIH Human Genome Project. In a brochure prepared for the meeting, conference organizers said that "genetic research holds out the prospect of identifying individuals who may be predisposed to certain kinds of criminal conduct."[29] When news of the planned meeting leaked out in the press, African American groups protested and called on the NIH to withdraw its funding for the gathering.

With nearly one out of every four young black men in America either in prison, on parole, or awaiting trial, African American leaders worried that such "scientific" efforts could lead to the genetic branding of blacks as biologically predisposed to violence and result in the imposition of new forms of social control under the cloak of "prevention."[30] U.S. Representative John Conyers, Jr., of Michigan, then chairman of the Congressional Black Caucus, said, "My worst fear is that they will unscientifically correlate race or social class with violent behavior. That would lead to discrimination policies."[31] Others, like Professor Ronald Walters, chairman of the Political Science Department at Howard University, expressed concern that government efforts to redefine violent crime as a "health" issue would lead to government programs to control young blacks with pacifying drugs, and made the analogy to the widespread use of Ritalin to pacify hyperactive schoolchildren.[32] By making violent crime a health problem, the public debate shifts from environmental factors that affect crime, like lack of educational opportunity, joblessness, and poverty, to genetic "errors" that can be controlled or weeded out.

Even as conference organizers and NIH officials were being vilified by critics in the public media, research on the genetic basis of violent crime was being conducted quietly in scientific facilities both in the United States and abroad. Scientists at the Bowman Gray School of Medicine in North Carolina have found that monkeys with lower levels of serotonin exhibit much more aggressive behavior than their peers and are "more likely to bite, slap or chase other monkeys."[33] They are also less social, spend more time alone, and have less intimate body contact with other monkeys. A similar study conducted by the National Institute on Alcohol Abuse and Alcoholism found that men with lower levels of serotonin are more likely to commit impulsive crimes, such as murdering strangers.[34]

A growing number of scientists are trying to link up "inborn" personality traits with a greater proclivity to commit violent crimes. The late Dr. Richard Herrnstein of Harvard University, whose theories on heredity and intelligence recently stirred controversy in a book he co-authored with Charles Murray entitled *The Bell Curve,* speculated that a genetic predisposition for "thrill seeking" and "restless impulsivity," and an accompanying inability to defer gratification, may be contributory factors in shaping a violent criminal.[35] Psychologist Dr. Jerome Kagan, also of Harvard University, believes that individuals with a high genetic threshold for anxiety and fear—individuals with lower than average heart rates and blood pressures—may be more "at risk" for violent criminal behavior. Kagan predicts that within twenty-five years, genetic tests may be able to identify fifteen children with violent tendencies out of every thousand. One of the fifteen children, says Kagan, will likely commit a violent crime sometime in the future.[36]

Research studies using PET scans—positron emission tomography—to track brain activity in violent patients by tracing radioactive substances have found specific types of abnormalities in their brains, suggesting a possible genetic basis for certain forms of violent behavior. In California, PET scans have been introduced into the courtroom and used as a scientific aid in the sentencing process to determine whether or not a convicted felon is likely to repeat his crime upon release. Legal scholars believe that the courts will increasingly rely on brain scans in determining sentencing and making parole decisions.[37]

## Genetic Discrimination

Societies have always been divided between the haves and the have-nots, the powerful and the powerless, the elite and the masses. Throughout history, people have been segregated by caste and class, with a myriad of rationales being used to justify the injustices imposed by the few on the many. Race, religion, and nationality are all well-worn methods of categorization and victimization. Now, with the emergence of genetic screening and genetic engineering, society entertains the prospect of a new and more serious form of segregation. One based on genotype.

A 1996 survey of genetic discrimination in the United States, conducted by Dr. Lisa N. Geller, et al., of the Department of Neurobiology and Division of Medical Ethics at Harvard Medical School, suggests that the practice is already far more widespread than previously thought. Genetic discrimination is being practiced by a range of institutions including insurance companies, health care providers, government agencies, adoption agencies, and schools. Researchers surveyed individuals at risk—or related to people at risk—for Huntington's disease, mucopolysaccharidosis (MPS), phenylketonuria (PKU), and hemochromatosis. Huntington's disease is a fatal disorder whose symptoms appear in middle age. MPS is associated with mental retardation and organomegaly. PKU also results in mental retardation but can be treated with special diets after birth. Hemochromatosis is an iron storage disorder.[38]

Of the 917 individuals surveyed in the study, 455 reported that they had experienced some form of discrimination based on their genetic makeup and genetic predispositions. Insurance companies and health providers were the most likely to practice genetic discrimination.[39] In one instance, a health maintenance organization refused to pay for occupational therapy after an individual was diagnosed with MPS-1, arguing that it was a preexisting condition. In another case, a twenty-four-year-old woman was refused life insurance because her family had a history of Huntington's disease, despite the fact that she had never been tested for the disease.[40] A *Newsweek* study of discrimination in the insurance industry found similar instances of abuse. One family had its entire coverage canceled when the insurance company discovered that one of its four children was afflicted with fragile X disease. The rest of the children were free of the disease but lost their coverage anyway.[41]

The growing incidence of genetic discrimination by insurance companies is all the more troubling because so many of the genetic diseases being screened for are capable of manifesting themselves in varying degrees of severity. There is often no way to know whether an individual will become symptomatic and, if so, when and to what degree. With some genetic diseases treatment is possible, but again, this can vary from person to person, with significant differences in results.

Discriminating against an individual for a preexisting genetic condition over which he or she has no control, or for a genetic predisposition that may never be expressed, or, if expressed, may be either non-serious or treatable, seems patently unfair. Many health professionals worry about millions of people being branded with lifelong "genetic scarlet letters," with consequences that extend far beyond the issuance of insurance policies.[42]

Nonetheless, insurance companies argue that if they are denied, by law, access to the accumulating body of genetic information, they might be forced out of business. A 1989 report prepared for the American Council of Life Insurance stated, "If insurers were unable to use genetic tests during the underwriting process because 'risk should only be classified on the basis of factors that people can control,' then equity would give way to equality (equal premiums regardless of risk) and private insurance as it is known today might well cease to exist."[43] More likely, insurance companies will simply pass on the additional costs of high-risk policy holders by increasing the premiums of low-risk policy holders. Such policies, in turn, will increase the burden on the poorest members of society, who will be unable to pay the higher premiums, and therefore be excluded from coverage.[44]

The costs of genetic screening tests will be a decisive factor in the decision to include genetic data in the assessment of insurance premiums and coverage. As long as current testing methods are relatively expensive, they are less likely to be widely used by care providers and insurance companies. As the cost of screening tests falls over the course of the next several years, the likelihood of widespread testing is all but a certainty.

Keeping medical records out of the hands of insurers is likely to be difficult. Still, twenty-six states have already passed laws to exclude insurers from discriminating on the basis of genetic predispositions. The insurance industry bitterly opposes the new laws and is actively lobbying in the state legislatures and in Congress to water down existing statutes and prohibit new laws from being enacted. Robert A. Meyer, a lawyer for the American Council of Life Insurance, says the industry "needs to be able to use genetic information, like all other medical information, in deciding whether or not to insure someone and at what price." Susan Van Gelder, a vice president of the Health Insurance Association of America, goes even further, saying, "We oppose any prohibition on the ability to collect information for risk assessment and risk selection." Many state legislators remain unimpressed by the industry's claims and argue that their constituents' fundamental rights to genetic privacy should prevail. J. Brian McCall, a Republican insurance executive and state legislator who authored the Texas law banning genetic discrimination by insurers, worries that "with genetic tests, insurance companies can virtually eliminate the guesswork in underwriting. They can seek out people who are genetically pure, creating a ghetto of the uninsured, because they will know who is likely to get a particular disease at a particular age."[45]

Couples desiring to adopt children have also been discriminated against because of their genotypes. In one reported instance, a couple was denied adoption because the wife was "at risk" of coming down with Huntington's disease. In another case a birth mother with Huntington's disease was not allowed to put up her child for adoption through a state adoption agency. Still another couple, one of whom was at risk of developing Huntington's, was denied adoption of a normal baby but was allowed to adopt a baby at risk of coming down with Huntington's disease.[46] The very idea that genetically "at risk" couples should only be paired with genetically "at risk" babies, and vice versa, is still another early warning sign of what might potentially develop into a kind of informal genetic caste system in the coming century.

Equally troubling is the prospect of discriminatory genotyping of racial and ethnic groups in the Biotech Century. As scientists gain more information on the workings of the human genome, they will succeed in identifying an increasing number of genetic traits and predispositions that are unique to specific ethnic and racial groups, opening the door to the possibility of genetic discrimination of entire peoples. Already, for example, we know that Armenians are more prone to Familial Mediterranean Fever disease. Jews are carriers of Tay-Sachs and Gaucher's disease, and Africans carry the sickle-cell gene.[47] Might not this kind of ethnic- and race-specific genetic information be used as well by institutions as a tool of discrimination, segregation, and abuse?

Employers are also becoming more interested in using genetic screening tests to screen prospective employees. Their reasons run the gamut. Some chemical companies are interested in screening workers to assess their genetic sensitivity to highly toxic work environments. Worried over the high cost of health insurance coverage, disability compensation claims, and absenteeism, companies are interested in weeding out workers who may be more susceptible to illness. Matching workers' genotypes to the workplace would be a less costly alternative to cleaning up the sites and making them safe for all the workers, regardless of their susceptibilities. Other employers who invest heavily in long-term education and on-site training want to know if prospective employees—especially those destined for advancement on the executive track—will be free of potentially debilitating diseases over the lifetime of their work contracts. With education and retraining costs increasing, employers are understandably anxious to ensure that they are not wasting both their time and resources on an employee who won't be able to continue working a few years down the line.

If a woman carries the BRCA-1 breast cancer gene, or a man carries the Huntington's gene or a genetic predisposition for early heart attack or stroke, corporate access to the information might well be a critical factor in hiring decisions and in promotion evaluations.

Some institutions whose employees require a high level of emotional stability, like defense contracting firms, airlines, and police departments, will no doubt be interested in genetic tests to detect predispositions for alcoholism, depression, and mood and behavioral disorders. They might well contend that the public is ill served if an air traffic controller, pilot, police officer, or safety engineer, suffers from mental illness—or might in the future. Similarly, the isolation of genetic traits and predispositions that condition personality traits, like thrill-seeking and extroversion, might be useful to companies employing test pilots and travel guides or salespersons and lobbyists.

The discovery of the sickle-cell anemia trait in the 1970s is a harbinger of the potential danger that can result from the unbridled use of genetic screening by employers. Carriers of the recessive gene—most of whom are African Americans—were denied entrance into the U.S. Air Force Academy for fear that they might suffer the sickling of their red blood cells in a reduced-oxygen environment. Others were denied jobs in toxic-sensitive environments in the chemical industry.[48]

The 1996 Geller survey on genetic discrimination also chronicled a growing number of employee discrimination cases. In one instance, a social worker was abruptly dismissed after her employer learned she was at risk of developing Huntington's disease. In the eight months preceding her termination, she had been promoted several times and received high performance evaluations. It should be noted that the fired employee was not symptomatic, nor had she ever been tested for the disease. Rather, she was dismissed because of knowledge that other family members had developed the illness.[49] She is not alone in being singled out for discrimination. In 1982, the federal Office of Technology Assessment conducted its own survey of American businesses and found that nearly half of the companies conducting screening tests had reported transferring or dismissing "at risk" workers.[50] A more extensive survey of four hundred employers conducted by the Northwestern Life Insurance Company in 1989 reported that 15 percent of the companies planned to conduct routine genetic screening tests of their prospective employees and dependents before the year 2000.[51]

Depriving prospective employees of jobs based on their genotypes runs the risk of creating a new group of dispossessed workers in the coming Biotech Century. With so much genetic information becoming available on every human being—from simple single-gene disorders to complex polygenic mood and behavior traits—it's likely inevitable that some employers will use the genetic data to select and sort out prospective employees and make more "informed decisions" about hiring and promotions. One's "genetic passport" is likely to play a sizable and, in some cases, a determining role in employment decisions in many industries in the future.

The notion of an army of genetically unemployable workers can no longer be as easily dismissed as it was just a decade or so ago. With health and disability compensation costs mounting each year in the United States, employers are going to feel increasingly pressured, by their own bottom lines, to genotype workers for jobs in hopes of cutting costs and increasing their profit margins.

To avoid the possibility of creating a genetically unemployable class, proper attention will need to be placed on setting limits and restricting institutions from practicing genetic discrimination. Laws will have to be established in every country to prohibit employers from discriminating against human beings whom they regard as "genetically less fit" for certain tasks based on predispositions that may or may not manifest themselves, and even if they do, may not seriously compromise an individual's ability to "get the job done." Concerned about the prospect of creating a "genetic underclass," the New York state legislature recently passed legislation that prohibits employers from discriminating against workers because of a "genetic predisposition or carrier status."[52]

Genetic discrimination is also beginning to extend to other institutions in society. Children are increasingly being classified, segregated, and treated in a discriminatory manner in the nation's schools based on only vague notions—and on occasion, misunderstandings—of the role played by genetic inheritance in academic and classroom performance.[53]

In the 1950s, '60s, '70s, and early '80s, academic success or failure was largely attributable to environmental factors, including parental nurturing, family dynamics, community support, and economic background.

Breakthroughs in cognitive psychology and molecular biology in the 1970s and 1980s set in motion a series of profound changes in the American education system with educators placing greater emphasis on biological, as opposed to environmental, causes of student performance. Writing in *The American Journal of Law and Medicine,* Dorothy Nelkin, professor of sociology and law at New York University, and Laurence Tancredi, a New York psychiatrist and lawyer, noted,

> Increasing research on human genetics has encouraged the assumption that learning and behavioral problems reflect biological deficits. Problems lie less in a student's environment and social situation than in the biological structure of his or her brain.[54]

Today, reading difficulties, short attention span, and behavioral problems are increasingly seen as biological deficits and classified as illnesses to be treated by pharmacological means and other forms of medical intervention. Nowhere is the shift in emphasis more apparent than in the reclassification of behavioral problems. In the 1960s and 1970s "hyperactivity" was considered a psychological and social problem and dealt with by attempting to understand the child and reform his or her environment. In 1980, the editors of the Diagnostic and Statistical Manual of Mental Disorders replaced the term "hyperactivity" with a new term, "attention deficit disorder," signaling a new belief that the problem lies in the brain chemistry and genetic endowment of the child and, therefore, should be treated as an illness. Millions of children with short attention spans, who exhibit hyperactive behavior, have been classified as suffering from attention deficit disorder and are now being treated for their illness with Ritalin and other drugs.[55]

Other behaviors have also been redefined in the medical literature, reflecting the new bias toward biological deficits as an explanatory model for student performance. A number of new biologically-based disabilities including "expressive writing disorders" and "stereotypy habit disorders" have found their way into the Diagnostic and Statistical Manual of Mental Disorders.[56] While it is only fair to acknowledge the genetic basis of many learning disabilities, educational reformers are concerned that the shift in emphasis to "genetic causation" has left little or no room for the obvious role played by the environment in shaping a child's mental abilities and behavioral orientation.

IQ tests, which purport to measure inherited intelligence, have been used for much of the twentieth century to classify, place, and track children, establishing a discriminatory scheme in the classroom. What's new is a generation of testing procedures designed to isolate and measure specific biological deficits in children for potential medical intervention. The Neural Efficiency Analyzer, for example, measures the speed with which the brain processes information. Brain electrical activity mapping (BEAM) is used by neurologists to diagnose students with learning disorders. With computer brain scans, neurologists can follow the flow of blood in the brain while a child is reading or thinking and detect differences in youngsters with dyslexia. As mentioned earlier, doctors using positron emission tomography (PET scans) can visually track brain activity by tracing radioactive substances. Researchers are using the PET technology to monitor brain activity in the frontal region of the brain and correlate it with childrens' behavior patterns in order to identify "at risk" youngsters.[57]

The array of new neurotechnologies gives further impetus to efforts in the schools to classify youngsters on the basis of their genotype. The shift to genetic determinism is occurring despite years of scientific studies showing that such things as diet, lifestyle, family, class background, and socializing experience also significantly affect learning skills, classroom performance, and personal behavior.

The increasing "genetization" of children has begun to undermine the traditional mentor relationship in the classroom, substituting the teacher-student dynamic with the practitioner-patient model. The evidence of this change is everywhere. Millions of school children are currently being treated with a range of pharmacological drugs for learning disabilities and behavioral disorders, including Librium and Valium for anxiety, Prozac and Zoloft for depression, Dexedrine for behavioral disorders, and phenobarbital and Benadryl for purposes of sedation.[58] Prescribed drugs have become a major factor in the education process and have been increasingly used to segregate and stigmatize children in the classroom.

Some teachers are less tolerant of children who have been diagnosed with a genetic disorder and more likely to give up on them, believing that their genetic "handicaps" make them less likely to keep up with "normal" children. Less attention, supervision, and support by teachers often translates into a downward learning spiral, diminution of personal confidence and self-esteem, and the further marginalization of genetically branded students. The discriminatory biases can follow children through school and into the workplace. Poorly educated and often improperly socialized, these youngsters are likely to be discriminated against a second time around by would-be employers for psychological and intellectual deficits created, in part, by the school systems which unfairly genotyped and stereotyped them in the first place.

Segregating individuals by their genetic makeup represents a fundamental shift in the exercise of power. In a society where the individual can be stereotyped by genotype, institutional power of all kinds becomes more absolute. At the same time, the increasing polarization of society into genetically "superior" and genetically "inferior" individuals and groups could create a new and powerful social dynamic. Those families who can afford to program "superior" genetic traits into their fetuses at conception could assure their offspring an even greater biological advantage, and thus a social and economic advantage as well. For the emerging "genetic underclass," the issue of genetic stereotyping is likely to lead to growing protests and the birth of a worldwide "genetic rights" movement as an increasing number of victims of genetic discrimination organize collectively to demand their right to participate freely and fully in the coming Biotech Century.

Some genetic engineers believe that a future genetocracy is all but inevitable. Molecular biologist Lee Silver of Princeton University writes about a not-too-distant future made up of two distinct biological classes, which he refers to as the Gen Rich and Naturals. The Gen Rich, which account for 10 percent of the population, have been enhanced with synthetic genes and have become the rulers of society. They include Gen Rich businessmen, musicians, artists, intellectuals, and athletes, each enhanced with specific synthetic genes to allow them to succeed in their respective fields in ways not even conceivable among those born of nature's lottery.

At the center of this new genetic aristocracy are the Gen Rich scientists, who are enhanced with special genetic traits that greatly increase their mental abilities, giving them the power to dictate the terms of future evolutionary advances on Earth. Silver says that:

> . . . with the passage of time, the genetic distance between Naturals and the Gen Rich has become greater and greater, and now there is little

> movement up from the Natural to the Gen Rich class. . . . All aspects of the economy, the media, the entertainment industry, and the knowledge industry are controlled by members of the Gen Rich class. . . . In contrast, Naturals work as low-paid service providers or as laborers. . . . Gen Rich and Natural children grow up and live in segregated social worlds where there is little chance for contact between them . . . [eventually] the Gen Rich class and the Natural class will become the Gen Rich humans and the natural humans—entirely separate species with no ability to cross-breed and with as much romantic interest in each other as a current human would have for a chimpanzee.[59]

Silver acknowledges that the increasing polarization of society into a Gen Rich and Natural class might be unfair, but he is quick to add that wealthy parents have always been able to provide all sorts of advantages for their children. "Anyone who accepts the right of affluent parents to provide their children with an expensive private school education cannot use unfairness as a reason for rejecting the use of reprogenetic technologies," argues Silver. Like many of his colleagues, Silver is a strong advocate of the new genetic technologies. "In a society that values human freedom above all else," writes Silver, "it is hard to find any legitimate basis for restricting the use of reprogenetics."[60]

## Difficult Choices

Even with all of the excitement being generated around the new genetic technologies, we sense, though dimly, the menacing outline of a eugenics shadow on the horizon. Still, we would find it exceedingly difficult to say no to a technological revolution that offers such an impressive array of benefits. Thus, we find ourselves on the horns of a dilemma as we make the first tentative moves into the Biotech Century. One part of us, our more ancient side, reels at the prospect of the further desacralizing of life, of reducing ourselves and all other sentient creatures to chemical codes to be manipulated for purely instrumental and utilitarian ends. Our other side, the one firmly entrenched in modernity, is zealously committed to bringing the biology of the planet in line with engineering standards, market forces, and progressive values. Not to proceed with this revolution is unthinkable, as it would violate the very spirit of progress, a spirit that knows no bounds in its restless search to wrest power from the natural world.

Finding new, more powerful technological ways of controlling and harnessing nature for utilitarian and commercial ends has been the ultimate dream and central motif of the modern age. It was Francis Bacon, the founder of modern science, who urged future generations to "squeeze," "mould," and "shape" nature, in order to "enlarge the bounds of human empire to the effecting of all things possible." Armed with the scientific method, Bacon was convinced that we had, at long last, a methodology that would allow us "the power to conquer and subdue" nature and to "shake her to her foundations."[61] Bacon laid the groundwork for the Enlightenment era that followed, by providing a systematic vision for humanity's final ascendance over nature. Isaac Newton, Rene Descartes, John Locke and other Enlightenment philosophers constructed a world view that continues to inspire many of today's molecular biologists and corporate entrepreneurs in their quest to capture and colonize the last frontier, the genetic commons that is the heart of the natural world.

Gaining access to and control over the very "blueprints" of life is the apotheosis of the Enlightenment vision. John Locke, the great political theorist of the Enlightenment, observed that human beings are, by nature, acquisitive and utilitarian. To negate nature, to remake it in our own image, to use it for our own ends, is to fulfill our destiny. Genetic technology represents the ultimate negation of nature. It confirms our long-held collective belief that all of nature has been put here for our exclusive use. To say no to the genetic engineering of life is to rob humanity of its rightful inheritance, its claim to sovereignty over the Creation. From its very inception, the modern age challenged the static medieval world view based on rigid adherence to custom and convention—what Voltaire satirically referred to as "the best of all possible worlds." Modernists preferred a new vision, "that all worlds are possible." The French aristocrat the Marquis de Condorcet excited the passions of an age when he confidently proclaimed:

> No bounds have been fixed to the improvement of the human faculties . . . the perfectibility of man is absolutely indefinite; . . . the progress of this perfectibility henceforth above the control of every power that would

> impede it, has no other limit than the duration of the globe upon which nature has placed us.[62]

E. O. Wilson, the Harvard professor and a founding father of the new school of sociobiology, takes Condorcet's vision to its ultimate conclusion. In the final passage of his Pulitzer Prize–winning book, *On Human Nature,* Wilson calls upon society to assume the role of architect of human evolution with the goal of perfecting the human species:

> In time, much knowledge concerning the genetic foundation of social behavior will accumulate, and techniques may become available for altering gene complexes by molecular engineering and rapid selection through cloning . . . The human species can change its own nature. What will it choose? Will it remain the same, teetering on a jerrybuilt foundation of partly obsolete Ice-Age adaptations? Or will it press on toward still higher intelligence and creativity, accompanied by a greater—or lesser—capacity for emotional response? New patterns of sociality could be installed in bits and pieces. It might be possible to imitate genetically the more nearly perfect nuclear family of the white-handed gibbon or the harmonious sisterhoods of the honeybees.[63]

The Biotech Century promises to complete the modernists' journey by "perfecting" both human nature and the rest of nature, all in the name of progress. Who could oppose the engineering of new plants and animals to feed a hungry world? Who could object to engineering new forms of biological energy to replace a dwindling reserve of fossil fuels? Who could protest the introduction of new microbes to eat up toxic wastes and other forms of chemical pollution? Who could refuse genetic surgery to eliminate crippling diseases?

The short-term benefits of the emerging biotechnical age appear so impressive that any talk of curtailing or preventing their widespread application is likely to be greeted with incredulity, if not outright hostility. How could anyone in good conscience oppose a technology that offers such hope for bettering the lot of humanity?

In the years to come, a multitude of new genetic engineering techniques will be forthcoming. Every one of the breakthroughs in biotechnology likely will be of benefit to someone, under some circumstance, somewhere in society. Each will, in some way, appear to advance the future security of an individual, a group, or society as a whole. The point that needs to be emphasized is that bioengineering is coming to us not as a threat but as a promise; not as a punishment but as a gift. While the thought of engineering living organisms still conjures up sinister images in the movies, it no longer does in the marketplace. Quite the contrary, what we see before our eyes are not monstrosities but useful products and hopeful futures. We no longer feel dread, but only elated expectations at the great possibilities that lie ahead for each of us in the Biotech Century.

For its most ardent supporters, engineering life to improve humanity's own prospects is, no doubt, seen as the highest expression of ethical behavior. Any resistance to the new technology is likely to be castigated by the growing legion of true believers as inhuman, irresponsible, morally reprehensible, and perhaps even criminal.

On the other hand, the new genetic engineering technologies raise one of the most troubling political questions in all of human history. To whom, in this new era, would we entrust the authority to decide what is a good gene that should be added to the gene pool and what is a bad gene that should be eliminated? Should we entrust the government with that authority? Corporations? University scientists? From this perspective, few of us would be able to point to any institution or group of individuals we would entrust with decisions of such import. If, however, we were asked whether we would sanction new biotech advances that could enhance the physical, emotional, and mental well-being of our progeny, many of us would not hesitate for a moment to add our support.

We appear caught between our instinctual distrust of the institutional forces that are quickly consolidating their power over the new genetic technologies and our desire to increase our own personal choices and options in the biological marketplace. While control of the new genetic technologies is being concentrated in the hands of scientists, transnational companies, government agencies, and other institutions, the products and services are being marketed under the guise of expanding freedom of choice for millions of consumers.

In the early stages of this new technological and commercial revolution, the informal bargain being struck between the governing institutions of society and consumers appears to be a reasonable one. Biotechnology has much to offer. But, as with the introduction of other technological innovations throughout history, the final costs have yet to be calculated. Granting power to a specific institution or group of individuals to determine a better-engineered crop or animal or a new human hormone seems a trifle in comparison with the potential returns. It is only when one considers the lifetime of the agreement that the full import of the politics of the biotechnological age becomes apparent.

Throughout history, some people have always controlled the futures of other people. Today, the ultimate exercise of power is within our grasp: the ability to control, at the most fundamental level, the future lives of unborn generations by engineering their biological life processes in advance, making them partial hostages of their own architecturally designed blueprints. I use the word "partial" because, like many others, I believe that environment is a major contributing factor in determining one's life course. It is also true, however, that one's genetic makeup plays a role in helping to shape one's destiny. Genetic engineering, then, represents the power of authorship, albeit limited authorship. Being able to engineer even minor changes in the physical and behavioral characteristics of future generations represents a new era in human history. Never before has such power over human life even been a possibility. Should power of this sort be granted to any public or commercial institution or, for that matter, even to consumers? Whether institutionally motivated or consumer driven, the power to determine the genetic destiny of millions of human beings yet to come lessens the opportunities of every new arrival to shape his or her own personal life story.

Still, at the dawn of the Biotech Century, the authorial power, though formidable, appears so far removed from any potential threat to individual human will as to be of little concern. Many of us will be eager to take advantage of the new gene therapies, both for ourselves and our offspring, if they deliver on their promise to enhance our physical, emotional, and mental health. After all, part of being truly human is the desire to alleviate suffering and enhance human potential. The problem is that biotechnology has a distinct beginning but no clear end. Cell by cell, tissue by tissue, organ by organ, we may willingly surrender our personhood in the marketplace. In the process, each loss will be compensated for with a perceived gain until there is little left to exchange. It is at that very point that the cost of our agreement will become apparent. But it is also at that point that we may no longer possess the very thing we were so anxious to enrich—our humanity. In the decades to come, we humans might well barter ourselves away, a gene at a time, in exchange for some measure of temporary well-being. In the end, the personal and collective security we fought so long and hard to preserve may well have been irreversibly compromised in pursuit of our own engineered perfection.

# Six Computing DNA

For more than a decade, futurists, economists, and policy makers have heralded the emergence of the information age. In their enthusiasm, they have lost sight of the central role communication technologies play in history and have mistakenly seen the computer and telecommunication revolutions as ends in themselves. Rather, like every great communications revolution, the information age technologies represent a powerful new vehicle for transforming the natural world and coordinating and managing the economic activity that flows from it. More than anything else, the computer is the communication tool and software, the language and text, for deciphering and utilizing the vast genetic resources of the Earth in the coming economic era. It is the sixth strand of the new operational matrix of the Biotech Century.

The coming together of the information sciences and life sciences—the computer and the gene—into a single technological and commercial revolution, presages the beginning of a new age in world history. In this regard, the computer's historic import is best understood in relation to the printing press, the communication technology that helped reshape our world, our economy, and our sense of self at the dawn of the industrial era.

## The Printed Page and the Industrial Age

The industrial age would not have been possible without the invention of the printing press by Johannes Gutenberg in the fifteenth century. The ability to mass-produce identical copies of an original written work, in a fraction of the time and with far less effort than went into medieval manuscript production, revolutionized Western society.

Print technology provided a new means of communications to manage the fast-paced, complex world of coal and steam power. Print also gave rise to a new way of organizing economic activity that would later be characterized as the industrial way of life. Equally important, the new communication technology fundamentally changed human consciousness itself, creating the new bourgeois man and woman of the modern era.

To begin with, print introduced the idea of assembly, a key component of the industrial way of life. Separating the alphabet into standardized units of type that were uniform, interchangeable, and reusable made print the first modern industrial process. Print also introduced the idea of rigid positioning of things in space. With print, objects are uniformly spaced by positioning type on a chase and locking the chase onto a press. The composite type can then be reproduced over and over, each copy identical and indistinguishable from the original. Assembly, uniform and interchangeable parts, predictable positioning of objects in space, and mass production were the foundation stones of the industrial way of life. Print created the archetype technology for this new way of organizing nature.

The new print medium also redefined the way human beings organize knowledge. The mnemonic redundancy of oral communication and the subjective eccentricities of medieval script were replaced by a more rational, calculating, analytical approach to knowledge. Print replaced human memory with tables of contents, pagination, footnotes, and indexes, freeing the human mind from continually recalling the past, so that it might fix on the present and future. The shift in consciousness prepared the way for the new commercial idea of unlimited material gain and human progress.

Print introduced charts, lists, graphs, and other visual aids that were to prove so important in creating ever more accurate descriptions of the world. Print made possible standardized, easily reproducible maps, making navigation and land travel more predictable and accessible. The opening up of oceans and land routes spread commercial markets and trade. Printed schedules, continually updated, mass-produced, and widely circulated, facilitated rail traffic and ocean voyages.

Print made possible a "contract" commercial culture by allowing merchants and capitalists to coordinate increasingly complex market activity and keep abreast of far-flung commercial transactions. Modern bookkeeping, schedules, bills of lading, invoicing, checks, and promissory notes were essential management tools in the organization of market capitalism. Print also made possible a uniform pricing system without which modern ideas of marketing and consumption could not have developed.

Print helped facilitate the development of nationalism and gave impetus to the creation of nation states. Vernacular languages, reduced to print, created a larger focus for collective identity. People began to see themselves as French, English, Germans, and Swedes. Print made possible detailed recordkeeping, so indispensable for the creation of modern government bureaucracies.

The print revolution helped create a new way of thinking, and with it, a new consciousness of how to act and interact with the world. Print organizes phenomena in an orderly, rational, and objective way, and in so doing, encouraged linear, sequential, and causal modes of thinking. The very notion of "composing" one's thoughts conjures up the idea of well-thought-out linear progression of ideas, one following the other in logical sequence, a mode of thought very different from that in oral culture where redundancy and discontinuity in conversation is often the rule.

By eliminating the redundancy of oral language and making precise measurement and description possible, print laid the foundation for the modern scientific world view. Phenomena could be rigorously examined, observed, and described and experiments could be made repeatable with exacting standards and protocols, something that was far more difficult to achieve in a manuscript or oral culture.

Print also made important the idea of "authorship." While individual authors were previously recognized, they were few in number. Manuscript writing was often anonymous and the result of the collective contribution of many scribes over long periods of time. It wasn't uncommon for several generations of scribes to lend their hand to a single written work. The notion of authorship elevated the individual to a unique status, separating him or her from the collective voice of the community.

The belief that one's ideas are "original" and "unique" helped foster the notion of the individual as a prime mover in the advance of civilization. The entrepreneurial spirit and competitive individualism, so championed by Adam Smith and the early Enlightenment thinkers, can also be traced, in part, to the new print revolution consciousness.

The idea of authorship also went hand in hand with the notion of owning one's own words. Copyright laws made communication between people a commodity for the first time. The idea that one could own thoughts and words and that others would have to pay to hear them marked a seminal turning point in the history of human relations. John Locke's idea of the commodity value of each person's ideas and labor in the marketplace—one of the critical tenets of modern capitalism—owes much to the change in consciousness brought on by the idea of personal authorship and copyright law.

Before print, people shared their thoughts together orally, in face-to-face dialogue and exchange. Even manuscripts were read aloud and were meant to be heard rather than seen. The print revolution helped nurture a more meditative environment. Books were read silently and alone, creating a new sense of personal privacy and along with it, notions of self-reflection and introspection, eventually leading to the creation of a therapeutic way of thinking about oneself and the world.

Finally, print made universal literacy possible for the first time, preparing successive generations with the communication tools they needed to manage the complexities of the modern market and new ways of working and socializing. In short, print created the appropriate mind-set and world view for an "industrious" way of living and being in the world.[1]

Print was grafted onto the late medieval feudal order and played a critical role in the success of the Protestant Reformation and the mercantile and urban revolutions that followed. Its destiny, however, lay with the industrial era. Print provided the all-important organizing language and mode of coordination for the steam and electricity revolutions, from which we fashioned the modern way of life. Even when later amplified by telegraphic and telephonic technologies and radio and television, print remained the dominant form of communication for organizing and coordinating the whole of the industrial age.

The symbiotic relationship between print technology, the steam engine, and the electrodynamo needs to be emphasized. The conversion of coal into steam, and later electricity, greatly accelerated both the speed and flow of economic activity. Previously, humanity had to rely on solar flow and on wind, water, and animal power for its sources of energy, setting upper limits to the amount of economic activity that could be generated. The new fast-paced world produced by the conversion of fossil fuels into raw energy could not be adequately managed by a slow-paced, face-to-face traditional oral culture or by means of handwritten script communications. Print technology provided a new means of communications whose essential features and attributes were sufficient to manage the first synthetic, high-energy, high-speed culture in history.

Today, print technology is being augmented and increasingly subsumed by computer technology in the organization and management of production, commerce, and trade. Although the computer and telecommunications revolutions are currently being grafted onto the industrial marketplace, the language and operating mode of information-based technologies are, in a real sense, at odds with many of the essential assumptions and governing principles of an older print culture and industrial way of life. That is one reason why, despite the vast sums of money spent on the new information technologies, productivity gains in the industrial economy have been less than expected. The new communication technologies' primary role will be to manage the genetic information of the new biotech marketplace just as print was used to manage an industrial marketplace built on fossil fuels.

## The New Language of Biology

Information technology is not so much an economic resource as it is a "language" of management and coordination. Its destiny is intimately linked to the raw genetic resources that it will isolate, download, organize, interpret, edit, and program in the coming Biotech Century. The computer and accompanying telecommunication technologies are, as the late media theorist Marshall McLuhan and others have pointed out, an extension of the human nervous system onto the world. They are a mechanical projection of the human mind into every nook and cranny of physical reality. At the same time, genes are the embodiment of biological existence, the myriad instructions organisms call upon to help orchestrate their life journey. Together, the computer and the gene create a powerful new mind/body dualism. In the coming years, humanity will come to use the computer more and more as a "substitute mind"—or language—to manipulate, redirect, and organize the vast genetic information that makes up the physical substance of living nature.

The computer organizes communications in a revolutionary new way that makes it an ideal tool for managing the dynamic flows and interactive processes that make up the fluid world of genes, cells, organs, organisms, and ecosystems. The letters and words are in the form of phosphorescent glows and are both ephemeral and frictionless. They do not exist *a priori* as individual solid units but rather come into existence on the screen when the software instructions call them forth. They have no past or future but exist only in the moment they are flickering on and off the screen. Electronic words are not locked into a fixed position like conventional print, but rather are dynamic, recombinable, easily edited, temporary and fleeting. In this sense they share much in common with the dynamic nature of genes, which continually flicker on and off in the process of writing and editing the script of life.

Electronic communication is organized cybernetically, not linearly. The notion of sequentiality and causality gives way to a total field of continuous integrated activity. In an electronic world of communications, subjects and objects give way to nodes and networks, and structure and function are subsumed by process. The computer's mode of organization—especially parallel computing—mirrors the processes of living systems where each of the parts is a node in a dynamic network of relationships that is continually readjusting and renewing itself at every level of its existence as it maintains a living presence.

Electronic communication also organizes knowledge differently than print technology. Hypertext replaces the more limited and narrow kind of print referencing. A self-contained book, with a set number of facts and references, gives way to an open-ended field of information, as footnotes and references are expanded indefinitely, creating new subtexts and meta-texts. Electronic communication is unbounded, open, integrative. On the other hand, it lacks the clarity that boundaries create. It is a form of communication that is closer to the flow of the unconscious mind, jumping from one thought to another, dwelling on parallel tracks, juxtaposing ideas one on top of another, straying off into unrelated or marginally related topics. It is less structured, less rational and analytic and more playful.

The electronic language of the computer is fashioning a new kind of human consciousness, one less Promethean and far more protean. The change in consciousness is attributable, in part, to the fact that electronic communication eliminates spatial context. The screen is not a physical place in the same sense as the printed page. Rather, it is a window into virtual worlds that transcends three-dimensional space altogether. It is an arena of fantasy where one is liberated from conventional boundaries and past and future constraints and free to create unlimited new environments in which to play.

In the virtual worlds behind the screen, the reality principle is steadily overtaken by the pleasure principle and fantasy gains the upper hand, subsuming the world of hard work and material accumulation in a new world made up of unlimited virtual experiences and personal transformations. Already, sociologists and psychologists like Sherry Turkle and Robert J. Lifton are observing a shift in consciousness among the first generation of computer-literate young people away from the older notion of the unique "centered self" to a new postmodern "multiple-persona"—a thespian consciousness, continuously being reedited, revised, recombined, and transformed to meet the ever-changing expectations of a more playful, complex, and multidimensional reality. All of these vast changes in consciousness and culture, brought on by the new electronic technologies and language, are laying the foundation for a new concept of nature and human nature as an unfinished work of art, continually reshapable into new forms and contexts.[2]

It is not only the computer's rules of engagement that make it a suitable communication tool to manage dynamic living systems. The very "operational language" of the computer is now being grafted onto biological systems. It is this common language that is creating a seamless web between the information and life sciences and making possible the joining together of computers and genes into a single, powerful technology revolution.

In 1953, just seven years after engineers plugged in and turned on the first working computer at the University of Pennsylvania in Philadelphia—the Electronic Numerical Integrator and Computer (ENIAC)—James Watson and Francis Crick announced that they had discovered the DNA double helix, unlocking the door to the secrets of the internal world of biology. As important as the discovery was the language they used to describe it. Borrowing metaphors and terms from the new field of cybernetics and the fledgling information sciences, Watson and Crick referred to the helix-like nature of the gene as a code, programmed with chemical information to be deciphered.

Joseph Weizenbaum of the Massachusetts Institute of Technology, an early pioneer in the computer field, notes that from the very beginning of the genetics revolution, the computer provided the overarching metaphor, and computer language provided the appropriate explanation, for understanding how biological processes function.

> The results announced by Crick and Watson fell on a soil already prepared by the public's vague understanding of computers, computer circuitry, and information theory. . . . Hence it was easy for the public to see the "cracking" of the genetic code as an unraveling of a computer program, and the discovery of the double helix structure of the DNA molecule as an explication of a computer's basic wiring diagram.[3]

Today, a half century later, information theory has become instrumental to deciphering and organizing as well as to understanding the increasingly complex world of molecular biology and genetic engineering.

In order to fully appreciate the extent to which the language of the computer has become the language of biology, it is necessary to delve into the operating principles that underlie the computer revolution. Those principles first took concrete form during World War II, when teams of engineers and scientists were assembled by the government with a mandate to devise new ways of organizing an increasing array of disparate information into an intelligent, efficient mode of operation. The undertaking was known as "operations research," and from it a new approach to organization emerged; it was called cybernetics, and it provided the operating principles for the computer revolution.

"Cybernetics" comes from the Greek word *kybernetes* which means "steersman." It is a general theory that attempts to explain how phenomena maintain themselves over time. Cybernetics reduces activity to two essential ingredients, information and feedback, and claims that all processes can be understood as amplifications of both.

Massachusetts Institute of Technology mathematician Norbert Wiener, the man who popularized cybernetic theory, defined information as the

> name for the content of what is exchanged with the outer world as we adjust to it, and make our adjustment felt upon it. The process of receiv-

> ing and of using information is the process of our adjusting to the contingencies of the outer environment, and of our living effectively within that environment.[4]

Information, then, consists of the countless messages and instructions that go back and forth between things and their environment. Cybernetics, in turn, is the theory of the way those messages or pieces of information interact with one another to produce predictable forms of action.

According to cybernetics theory, the "steering" mechanism that regulates all behavior is feedback. Anyone who has ever adjusted a thermostat is familiar with how feedback works. The thermostat monitors temperature in a room. If the room cools off and the temperature dips below the mark set on the dial, the thermostat kicks on the furnace, and the furnace remains on until room temperature coincides once again with the temperature set on the dial. Then the thermostat turns off the furnace, until the room temperature drops again, requiring additional heat. This is an example of negative feedback. All systems maintain themselves by the use of negative feedback. Its opposite, positive feedback, produces results of a very different kind. In positive feedback, a change in activity feeds on itself, reinforcing and intensifying the process, rather than readjusting and dampening it.

Cybernetics is primarily concerned with negative feedback. Wiener points out, "For any machine subject to a varied external environment to act effectively it is necessary that information concerning the results of its own action be furnished to it as part of the information on which it must continue to act."[5] Feedback provides information to the machine on its actual performance, which is then measured against the expected performance. The information allows the machine to adjust its activity accordingly, in order to close the gap between what is expected of it and how it in fact behaves. Cybernetics is the theory of how machines regulate themselves in changing environments. But, more than that, cybernetics is the theory that explains purposeful behavior in machines.

Wiener first introduced the notion that machines can exhibit purposefulness. In a landmark article published in *The Philosophy of Science* in 1943, Wiener defined purposeful behavior as "a final condition in which the behaving object reaches a definite correlation in time or space with respect to another object or event."[6] For Wiener, all purposeful behavior reduces itself to "information processing." He wrote,

> It becomes plausible that information . . . belongs among the great concepts of science such as matter, energy and electric charge. Our adjustment to the world around us depends upon the informational windows that our senses provide.[7]

After careful deliberation, Wiener concluded that "society can only be understood through a study of the messages and the communications facilities which belong to it."[8] Wiener came to view cybernetics as both a unifying theory and a methodological tool for reorganizing the entire world. Succeeding generations of scientists and engineers concurred. With the aid of the computer, cybernetics has become the primary methodological approach for organizing economic and social activity. Virtually every activity of importance in today's society is being brought under the control of cybernetic principles. "Information processing" by means of the computer is fast becoming the hallmark of our technological culture. Nowhere is this more in evidence than in our economic system. Once considered an adjunct to the management of large-scale economic organizations, information processing now permeates every aspect of the corporate structure and defines the organization itself. Corporations are increasingly being thought of as information systems embedded in networks of relationships.

Cybernetics has not only changed the way we go about organizing the world, but it also has affected the way we go about conceptualizing it. To begin with, the operating assumptions of cybernetics are antithetical to the orthodox view of the relationship between parts and whole. During the industrial era, it was assumed that the whole was merely an aggregate of the assembled parts that made it up. Cybernetics, in contrast, views the whole as an integrated system. The constant feedback of new information from the environment and the continual readjustment of the system to the environment set up a circular process in contradistinction to the linear mode of organization that characterized the Industrial Age. The self-correcting circularity of this new mode of organization "blurs the distinction between cause and effect."[9] According to the information theorists, in an increasingly complex environment it is no longer possible to entertain the simple fiction that one event in isolation leads to another event in isolation. We are now coming to realize that every event in some way affects everything else. Because everything is interrelated, it is necessary to organize activity into integrated systems.

Charles R. Dechert, professor of political science at the Catholic University of America, sums up the importance of the new set of organizing principles that have replaced the "assembly line" mentality of the Industrial Age: "Cybernetics extends the circle of processes which can be controlled—this is its special property and merit."[10] Increased reliance on the electronic computer ensures the institutionalization of cybernetic principles as the central organizing mode of the future.

It's worth noting, at this point, that without electricity, cybernetics and the computer revolution would have been impossible. Electricity, observed Marshall McLuhan, provided the means for the "instant synchronization of numerous operations," and by so doing, it "ended the old mechanical pattern of setting up operations in lineal sequence."[11] Electricity allows us to greatly enhance our capacity to anticipate and adjust to increasing levels of activity in the surrounding environment.

All earlier modes of technology were necessarily "partial and fragmentary" because they were unable to overcome their spatial context. Electricity allows humanity to skip over long stretches of space virtually instantaneously. McLuhan was one of the first to observe that with electric media humanity is able for the first time to "abolish the spatial dimension."[12] As a result, says McLuhan, "what emerges is a total field of inclusive awareness."[13] By subsuming space, electricity redefines all activity as pure process.

When Norbert Wiener published the first edition of his book *Cybernetics,* he included a subtitle: "Control and Communication in the Animal and the Machine." Wiener was convinced that the operating principles of cybernetics could be successfully extended from the engineering field to the life sciences. His goal was to reformulate biology in engineering terms, making it subject to rigorous mathematical analysis. Wiener once remarked that the only real difference between fire control in anti-aircraft gunnery and biological processes was the degree of complexity governing their respective information-sorting and feedback capacities.

Philosophy of science professor emeritus Marjorie Grene, of the University of California at Davis, suggests that as Wiener and other engineers improved the capacity of machines to regulate their own performances, they became more convinced that their own handiwork behaved much like living systems. Ergo, they concluded that the same operational guidelines they were imposing upon their technology must bear some correlation with the operating guidelines that animate living systems. According to Grene, a new way of thinking began to take hold in both the engineering field and in the life sciences.

> It says in effect: look to engineering, to blueprints and operational principles . . . for the source of your theoretical models in biology, much as Darwin drew on the works of sheep breeders and pigeon fanciers as a source for Natural Selection.[14]

Throughout the 1950s, engineering terminology continued to find its way into the field of biology, successfully transposing much of the language of the life sciences into a machine idiom. For example, consider the term "performance." Engineers use this word to refer to the activity of machines. Biologists, in contrast, have traditionally relied on the word "behavior" when referring to the activity of living organisms. Performance conjures up the idea of purposeful activity designed to meet a specific utilitarian objective. Behavior, on the other hand, often connotes the image of unpredictable and undirected activity without specific goals. Experimental psychologist R. L. Gregory of Cambridge University notes that biologists began to use the two words interchangeably, "the reason almost certainly being the influence of cybernetic ideas, which have unified certain aspects of biology and engineering."[15] Gregory goes on to say that the term "performance" was being relied on increasingly as biologists came to redefine living organisms in terms of relative efficiencies. Clearly the engineering mentality had permeated the field of biology; so much so that, as zoologist William H. Thorpe and psychologist Oliver L. Zangwill of Cambridge University, point out, living organisms were more and more being described in terms of their thermal efficiency, information efficiency, capital costs, running cost, and other technologically conceived criteria.

Wiener dreamed of unifying engineering and biology, and apparently many technicians in both fields were anxious to share his vision. In their book *Current Problems in Animal Behavior,* Thorpe and Zangwill assessed the impact information theory had exerted on the field of biology by the early 1960s and concluded that the life sciences were already succumbing to the operating assumptions of the cyberneticians. The two scholars noted, "Principles derived from control and communications engineering are being increasingly brought to bear upon biological problems and models derived from these principles are proving fertile in the explanation of behavior."[16] According to Thorpe and Zangwill, scientists in both fields were already finding common ground "under Norbert Wiener's banner of Cybernetics."[17]

More biologists were coming to view living organisms as information systems. Thorpe defines living organisms as things that "absorb and store information, change their behavior as a result of that information, and . . . have special organs for detecting, sorting and organizing this information."[18] The older Newtonian model, which viewed nature as "the movement of a particle under the action of a force," had been replaced with a new model that defines nature as "the storage . . . and the transmission of information within a system."[19] When one stops to consider that information is nonmaterial, the full impact of the revolution in thinking begins to come into focus. Because it is nonmaterial, information does not exist in a static spatial context in the manner that Newton had in mind when he defined the world in terms of matter in motion. When a biologist talks about living organisms as information systems, he or she is saying that they are instructions or programs that "describe a process and further, instruct that this process should be done."[20] When a biologist talks about process, he or she is referring to something that takes place over a period of time. Therefore, living systems in the new way of thinking are information programs that unfold in a predictable manner over time. "The most important biological discovery of recent years," says Thorpe, "is the discovery that the processes of life are directed by programmes . . . [and] that life is not merely programmed activity but self-programmed activity."[21]

Biologist Richard Dawkins goes even further, suggesting that we should no longer think of life as "fire," "sparks," or "breath" but, rather, as "a billion discrete digital characters carved in tablets of crystal." He writes,

> [Molecules of living things] are put together in much more complicated patterns than the molecules of non-living things, and this putting to-

> gether is done following programs, sets of instructions for how to develop, which the organisms carry around inside themselves. Maybe they do vibrate and throb and pulsate with "irritability" and glow with "living" warmth, but these properties all emerge incidentally. What lies at the heart of every living thing is not a fire, not warm breath, not a "spark of life." It is information, words, instructions . . . If you want to understand life, don't think about vibrant, throbbing gels and oozes, think about information technology.[22]

In his 1985 book *The Origins of Life,* physicist Freeman Dyson brought together the information and life sciences into a simple conceptual framework with this pithy observation:

> Hardware processes information; software embodies information. These two components have their exact analogues in the living cell; protein is hardware and nucleic acid is software.[23]

The French biologist Pierre Grassé has laid out a detailed presentation of the new approach to the conceptualization of nature, using the language of cybernetics. Grassé begins by framing all of life in cybernetic terms: "Information forms and animates the living organism. Evolution is, in the end, the process by which the creature modifies its information and acquires other information."[24]

Grassé goes on to develop a cybernetic model of a living organism. He starts with the strands of DNA that make up the genetic code. According to Grassé, the code represents the intelligence of the species. Grassé is willing to concede that DNA is ". . . the depository and distributor of the information," but he takes exception to James Watson, Francis Crick, and many of the neo-Darwinists who contend that it is also the "sole creator."[25]

Grassé likens the genetic code of an organism to a library and argues that neither one fabricates information; they are merely repositories of information. Both DNA and the library classify and store. It is at this point that Grassé applies the principle of "feedback" to living systems.

> DNA has to receive messages either from other parts of the cell or from organs . . . or from the outside world (sense stimuli, pheromones, etc.). Of itself, by what miracle could it generate information adequate to performance of a given function?[26]

To illustrate his point, Grassé uses the computer as an appropriate metaphor.

> The computer is limited in its operations by the program controlling it and the units of information fed into it. To enlarge its possibilities, its contents have to be enriched. What is new comes from outside.[27]

Grassé concludes that the living organism, like the computer, has "to be programmed and fed with external information in order for novelties to emerge."[28] The picture he sketches is a cybernetic model of life; a circular process in which the genes, the organism, and the environment continually feed information back and forth, allowing the organism to regulate itself in response to changing external cues. Grassé's more expansive view of the cybernetic relationship between genes, organism, and environment is a far cry from the simple genetic reductionism of Watson, Crick, and other earlier molecular biologists who, while they used the language of cybernetics and information theory to explain biological phenomena, continued to view the gene as "the master" molecule that sets in motion the organic process.

In the fifty years that have elapsed since Norbert Wiener sketched out his grand design, biological thinking has been recast in the image of information technology. By the late 1980s molecular biology textbooks had been virtually rewritten to reflect the influence of the information sciences. In one of the most widely used textbooks, *The Molecular Biology of the Cell,* the authors state,

> For cells as for computers, memory makes complex programs possible; and many cells together, each one stepping through its complex developmental control program, generates a complex adult body . . . thus the cells of the embryo can be likened to an array of computers operating in parallel and exchanging information with one another.[29]

Computer scientist Richard M. Karp, of the University of Washington in Seattle, one of a growing number of experts in the information sciences who is working alongside molecular biologists to manage genetic information, echoes the opinions of many of his colleagues in both fields when he says, "Biology, particularly at the molecular level, can be viewed for many purposes as an information science. To understand the cell, the brain, or the immune system, you sometimes have to view it as a very complex information processing system."[30]

Establishing cybernetics and information theory as a common language for both the computer sciences and the life sciences provides the all-important communication framework for using computers to manipulate and organize the vast genomic data flow of the coming Biotech Century.

Dr. Evelyn Fox Keller, professor of history and philosophy of science at the Massachusetts Institute of Technology, captures the significance of the epoch-making marriage of the information and life sciences by placing it within the context of the broader playing field being readied for the coming century. She writes,

> The body of modern biology, like the DNA molecule—and also like the modern corporation or political body—has become just another part of an informational network, now machine, now message, always ready for exchange, each for the other.[31]

## The Marriage of Computers and Genes

Today, molecular biologists around the world are busily engaged in the most extensive data collection project in history. In government, university, and corporate laboratories, researchers are mapping and sequencing the entire genomes of creatures from the lowliest bacteria to human beings with the goal of finding new ways of harnessing and exploiting genetic information for economic purposes. By the end of the twenty-first century, molecular biologists hope to have downloaded and catalogued the genomes of tens of thousands of living organisms—a vast library containing the evolutionary "blueprints" of many of the microorganisms, plants, and animals that populate the Earth. Mapping the genomes of so many species "will yield quantities of information that will dwarf by orders of magnitude anything encountered before," says biochemist Charles Cantor, the chief scientist at the Department of Energy's Human Genome Project.[32] The biological information being generated is so great that it can only be managed by computers and stored electronically in thousands of databases around the world. For example, the complete human sequence—only one of millions of species that will be sequenced and mapped—were it to be typed out in the form used in a telephone directory, would take up two hundred volumes of New York City's borough of Manhattan thousand-page directory. That's a database containing more than 3 billion entries. Taking the analogy a step further, if we were to print out the data on all human diversity, the database would be at least four orders of magnitude bigger—or ten thousand times the size of the first database. In the future, scientists are likely to concentrate their efforts on micromanaging and updating the databases for small regions of the genome of individual species, and coordinating their research with others by way of "genome work stations"—computer terminals that can provide researchers with access to the genomic databases of their colleagues around the world.[33]

Collecting, downloading, managing, and utilizing genomic information will require closer cooperation between the information and life sciences and the cross-training of researchers in the related fields of physics, mathematics, engineering, computer science, chemistry, and molecular biology. The Human Genome Project has hastened the coming together of the computer and genetic sciences. Sequencing and analyzing the three billion base pairs would not be possible without the help of computer scientists and increasingly sophisticated computational techniques. Richard Karp observes that the Human Genome Project is

> turning biology into an information science. Many biologists consider the acquisition of sequencing to be boring. But from a computer science point of view, these are first-rate and challenging algorithmic questions.[34]

Commenting on the Human Genome Project and bioinformatics, an analyst for the Association for Computing Machinery made the very prescient observation that

> In addition to [the] . . . genetic code [being self-modifying], biology also involves concepts that are dynamic, or fluid, meaning that the phenomena under study and the scientists' understanding of them keep changing. Creating a biological database that can accommodate concepts as their definitions evolve and that allows for experimental error presents major technical challenges not encountered in developing commercial or engineering databases.[35]

Mapping and sequencing the genomes is just the beginning. Reorganizing the whole of the natural world at the genetic level, with an eye to converting it to an array of useful commodities in the marketplace, is a daunting challenge, and easily the greatest management task ever conceived. Understanding and chronicling all of the webs of relationships between genes, tissues, organs, organisms, and external environments, and the perturbations that trigger genetic mutations and phenotypical responses, is so far beyond any kind of complex system ever modeled that only an interdisciplinary approach, leaning heavily on the computational skills of the information scientists, can hope to accomplish the task.

The potential power of the computer to decipher and manage genes became apparent in 1983 when Russell Doolittle, then professor of chemistry at the University of California at San Diego, and colleagues were able to make a significant biological discovery by merely reading computer printouts. The Doolittle research team compared computer printouts for two proteins and found that the DNA sequence in a type of cancer was also the same as a DNA sequence in a cellular growth, revealing that cancer genes created abnormal growth in cells. The discovery came about without doing a single biological experiment.

Robert Cook-Deegan, currently the director of the National Cancer Policy Board of the National Academy of Sciences, notes that beginning in the 1980s,

> the intrusion of computers into molecular biology shifted power into the hands of those with mathematical aptitudes and computer savvy . . . A new breed of scientist began to rise through the ranks, with expertise in molecular biology, computers, and mathematical analysis.[36]

Not surprisingly, bioinformatics has suddenly come of age. Titans in the computer field like Bill Gates, and Wall Street insiders like Michael Milken, are pouring funds into the new field of bioinformatics, in hopes of advancing the marital partnership of the information and life sciences. One of the pioneers in the new field is Leroy Hood, a biologist originally from the California Institute of Technology, who is often referred to in the trade as biotechnology's "premier tool-and-die man." Hood was one of the co-inventors of the gene sequencing machine that is used by researchers to identify the sequence of the 3 billion molecules that make up the human genome. The automated machine is sixty times faster than the older manual methods of sequencing. Milken has donated $25 million to Hood's research in hopes that his team of engineers and biologists will be able to build faster machines in order to locate the genetic causes of prostate cancer, from which Milken suffers. Gates has also invested heavily in Hood's firm, in the hope that he will do for genetic engineering what Henry Ford did for automobile production—speed up and automate the biotech production line.[37]

Hood's company, Darwin Molecular Corporation, headquartered in Bethell, Washington, is applying state-of-the-art computing technology to the task of deciphering the potentially valuable information coded in the genetic pool of the planet. Hood's firm, however, is not alone. Genomics research has become the hot new field in molecular biology and is bringing together experts from both the information sciences and life sciences to locate, map, sequence, analyze, download, catalog, and model genomic information across the plant and animal kingdom.

Companies with names like Bio-Image, Textco, Biosoft, the Oxford Molecular Group, Applied Biosystems, and Pangea Systems are cementing the marriage between these two great technology revolutions with the development of sophisticated software packages and programs designed to read, interpret, and manage genomic data. Bio-Image's DNA Sequence Film Reader and Sequence Assembly Manager Software Packages "automatically call bases, resolve ambiguities, align sequences, and assemble contigs" (piece together overlapping sequence segments end-to-end). Bob Luton, project manager for Bio-Image, says, "The software is smart enough to think."[38] Textco's Gene Inspector software package combines a sequence analysis package and an electronic laboratory notebook that "allows biologists to track, edit, and update their analysis."[39] Joel Bellenson, CEO of Pangea Systems, says that while the new breed of genomic firms are, for now, concentrating their efforts at the level of the individual molecule, in the future he envisions molecular biologists and computer engineers moving beyond DNA and protein sequence homology searching and into signaling and metabolic pathways—in other words, beginning to explore the more complex questions of how molecules interact with one another. "Instead of having this very individualistic molecular focus, we'll begin looking at the cell as an ecosystem of molecules."[40]

Bill Gates sums up the new collaborative efforts between the information and life sciences by saying, "This is the information age, and biological information is probably the most interesting information we are deciphering and trying to decide to change. It's all a question of how, not if."[41]

In an effort to better coordinate the flood of information generated by the various genome mapping projects, the National Institutes of Health (NIH) has established the National Center for Biotechnology Information at its Bethesda, Maryland, campus. The goal of the NIH project is to create an integrated genetic database that can be downloaded efficiently by researchers all over the world. A similar project, the European Bioinformatics Institute, has been established near Cambridge, England. To effectively integrate all the separate genetic databases into a centralized system, experts in the fields of bioinformatics are experimenting with creating a universal language that could be used to exchange all of the data among researchers.[42]

Computers are also being used to create virtual biological environments from which to model complex biological organisms, networks and ecosystems. The virtual environments help researchers create new hypotheses and scenarios that will later be used in the laboratory to test new agricultural and pharmaceutical products and medical treatments on living organisms. Working in virtual worlds, biologists can create new synthetic molecules with a few keystrokes, bypassing the often laborious process—which can take years—of attempting to synthesize a real molecule on the lab bench. With three-dimensional computer models, researchers can play with various combinations, on the screen, connecting different molecules to see how they interact. In 1991, research teams at Pennsylvania State University and the Scripps Clinic in La Jolla, California, created the first synthetic molecule with valuable chemical properties, using state-of-the-art computing capabilities. The compound, known as QM212, was conceived on a computer screen and its real-life counterpart is now being batch-produced in several biotech laboratories. Scientists plan to create all sorts of new molecules in the future using the new information-age computing technologies. Chemists are already talking about "compounds that could reproduce themselves, conduct electricity, detect pollution, stop tumors, counter the effects of cocaine, and block the progress of AIDS."[43]

In 1996, the molecular biology community stunned Wall Street with the announcement of the first DNA chip. As mentioned in Chapter One, the chips, which closely resemble computer chips, are packed with DNA and are designed to "read" the reams of genetic information in the genomes of living organisms. Affymetrix, a biotech start-up company in Santa Clara, California, designed the chips to detect genetic abnormalities. The company is currently working with OncorMed, a genetics testing firm in Gaithersburg, Maryland, to develop a DNA chip that can detect "P53" gene malfunctions, considered to be a contributing factor in more than 60 percent of all human cancers. Scientists say the day is not far off when DNA chips will be able to scan an individual patient, read his or her genetic makeup in precise detail, and even be able to detect abnormal or malfunctioning genes. DNA chips will eventually be able to determine which genes are flicking "on" and "off" at any given time. Dr. Mark Schee, of Affymetrix, says, "We can contemplate measuring the expression of every gene, like before and after someone drinks a cup of coffee." Other DNA chips might be used to scan a throat swab to identify a specific microbe that might be the cause of a patient's sore throat, even identifying specific genes in the bacteria that are resistant to certain antibiotics.[44]

DNA chips are made by using photolithography, the same technique used to make microprocessors. And like their predecessor, DNA chips are becoming ever more information rich. Affymetrix's first prototype in 1994 held 20,000 DNA probes. Today, its chips contain more than 400,000 probes.[45]

The final integration of the information and life sciences comes in the form of the "molecular computer," a thinking machine made of DNA strands rather than silicon. Scientists have already constructed the first DNA computer, and a growing number of both computer scientists and molecular biologists predict that sometime in the early years of the Biotech Century, much computing will take place along DNA pathways rather than on the integrated circuitry of a microchip. DNA's ability to compute information greatly exceeds the most advanced supercomputers that exist. Unlike most conventional computers, which are sequential and can only handle one thing at a time, DNA is a massive parallel computing machine and can theoretically compute a hundred million billion things at once.[46] One scientist recently quipped that a small jug of DNA can compute more arithmetic than all of the computers currently in use.[47]

The molecular computer is the brainchild of Dr. Leonard Adelman, a mathematician and professor of computer science at the University of

Southern California. Realizing that DNA stores information in much the same way as computers, Adelman hit on the idea that "you could use DNA to compute." Adelman points out, "DNA is essentially digital. This means it can count."[48] The question, then, is how to make it solve problems. In 1994, Adelman accomplished this task, getting DNA to solve a simple mathematical puzzle. The results, published in the journal *Science,* were greeted by the research community as a breakthrough experiment in the history of science.

Richard Lipton, a computer scientist at Princeton University, took Adelman's work a step further, by inventing a coding procedure for translating DNA base pairs into strings of ones and zeroes. He then poured together the contents of test tubes filled with genetically sequenced molecules, which allowed the DNA "to simulate the electronic gates by which computers make their yes-no decisions."[49] In short, he got the DNA to "think." Lipton believes that a DNA supercomputer could fit into a "bathtub" and would probably cost less than $100,000. Adelman says that DNA supercomputers will run a million times faster than today's most advanced supercomputer, and fundamentally change the way we live and the kind of world we live in. The DNA supercomputer brings the information sciences and life sciences together into a single technology revolution, with the power to remake the world.[50]

# Seven Reinventing Nature

Every major economic and social revolution in history has been accompanied by a new explanation of the creation of life and the workings of nature. The new concept of nature is always the most important strand of the matrix that makes up any new social order. In each instance, the new cosmology serves to justify the rightness and inevitability of the new way human beings are organizing their world by suggesting that nature itself is organized along similar lines. Thus, every society can feel comfortable that the way it is conducting its activities is compatible with the natural order of things and, therefore, a legitimate reflection of nature's grand design.

For more than a century, our ideas about nature, human nature, and the meaning of existence have reflected the extraordinary influence of Charles Darwin's theory of the origin and development of species. It would be difficult for most of us to imagine a world without his theory to inform and guide our journey. Now, however, this pillar of twentieth-century thought is being shaken from its foundation. Our ideas about nature, evolution, and the meaning of life are being fundamentally revamped as we enter the Biotech Century. As mentioned in Chapter Six, even the language and text we use to describe the evolutionary process is being rewritten. The new ideas about nature that are emerging will likely reshape our consciousness, values, and culture as significantly as did Darwin's theory of evolution when it replaced the God-centered creationist view of Christianity more than one hundred years ago.

Darwin constructed a theory of nature that, in its every particular, reinforced the operating assumptions of the Industrial Age. In so doing, he provided something much more valuable than a mere theory of nature. Darwin gave industrial man and woman the assurance they needed to prevail against any nagging doubts they might have regarding the correctness of their behavior. His theory confirmed what they so anxiously wanted to believe: that the way they were organizing their existence was "harmonious" with the natural order of things. Similarly, our newest ideas about evolution seem to be compatible with the new way we are going about organizing economic life in the Biotech Century, once again providing a much needed assurance that what we are doing is a mere reflection of the natural order of things and therefore both justifiable and inevitable. The new cosmological narrative represents the seventh strand of the operating matrix of the Biotech Century.

The role cosmology plays in rationalizing the new economic circumstances society finds itself in is critical. It is the least considered, yet most important feature of any new governing matrix and the linchpin upon which the entire edifice rests. It should be noted that once a new cosmology is widely accepted, the chances of generating a thoughtful debate over the way the economy and society have been reorganized is slim, as the public at large has already come to see the new economic and social reorganization as an amplification of, rather than a deviation from, nature's own operating assumptions. Any criticism, therefore, is likely to be regarded as suspect, as it would appear to fly in the face of a social order that is organized, in its every detail, to reflect the natural order.

Concepts of nature always focus on the big questions: Where did we come from? Why are we here? Where are we headed? For as long as we have had a history, human beings have had, at their disposal, a set of readily available answers as to what nature and life are all about. Where do these answers come from? How reliable are they? Why do answers we have long assumed to be beyond reproach suddenly become objects of ridicule and scorn? Are the new answers that replace them any more valid or are they ultimately doomed to the same ignominious fate?

The fact is, we human beings cannot live without some agreed-upon idea of what nature and life are all about. When we ponder what our own personal fate might be after the last breath of life is spent, or when we try to imagine what existed before existence itself, we are likely to become paralyzed with doubt. Our concept of nature allows us to overcome these ultimate anxieties. It provides us with some of the answers, enough to get along. A concept of nature, then, is more than just an explanation of how living things interact with one another. It also serves as a reference point for deciphering the meaning of existence itself. More than that, concepts of nature are the critical social constructs by which every society measures itself and justifies its relationship to the surrounding world.

Otto Rank, one of the great psychoanalysts of the twentieth century, suggests that our concepts of nature are supremely self-serving, reflecting our desire to make everything conform to our current image of ourselves. Rank believes that our concepts of nature tell us more about ourselves at any given moment of time than they do about nature itself. Historian of science Robert Young of Cambridge University would agree with Rank. He argues that there is no neutral naturalism. When we penetrate to the core of our scientific beliefs, says Young, we find that they are as much influenced by our culture as all our other belief systems. More to the point, anthropologist C. R. Hallpike of McMaster University in Canada contends that "the kinds of representation of nature . . . that we construct" flow from the way "we interact with the physical environment and our fellows."[1] In short, our concepts of nature are utterly, unabashedly, almost embarrassingly anthropocentric.

Try to imagine a society faithfully adhering to a concept of nature that is at odds with the way it goes about structuring its day-to-day activity. Obviously, a concept of nature must be compatible with the way people behave within a given cultural milieu if it is to both make sense and be acceptable. This has always been the case.

Every civilization justifies its behavior by claiming to have the natural order on its side. In each case, the process of legitimization is the same. A society organizes itself and its environment. Hierarchies are set up. Relationships are determined. Tasks are allocated. Rewards are distributed. But how do the members of society know that the way they've set up their society is the right way? This is the ultimate political question that every society faces. The answer amounts to a conjurer's sleight of hand. Since a society's view of what the whole world is all about is heavily influenced by the way it is organizing its own immediate world each day, it is only "natural" for the culture to come to the conclusion that the economic, political, and social reality it feels and experiences must, in fact, be reality. Therefore, it is only a short jump to fashioning a model of nature that is strikingly similar to the world being fashioned by the society. Then, not surprisingly, people find that their behavior does, in fact, correspond to the order of nature and, for that reason, conclude that the existing social order is appropriate. What better legitimization can there be for any governing body? Individuals rule and institutions prevail as long as enough people remain convinced that such behavior is merely a reflection of "the natural order of things."

Concepts of nature also serve as essential political instruments for eliciting unequivocal "deference and resignation." No one in his right mind would suggest that it is correct or even possible to resist the natural order. And if society happens to be unjust, exploitive, repressive, what is a person to do? If it's merely a reflection of the natural order of things, or at least structured in a way that adheres to nature's grand design, then to challenge it in any fundamental way would be as foolhardy and self-defeating as challenging nature itself. For society at large, and for ruling elites in particular, a concept of nature provides a mantle of legitimacy for the existing social order.

In the late medieval era, most Europeans accepted the official Church view of the origin of species spelled out by the great medieval churchman St. Thomas Aquinas in the thirteenth century. St. Thomas borrowed heavily from Hebraic and Hellenic thought, adding some of his own ideas in the process. The result was a cosmology that legitimized the existing social order while relieving the powers that be of responsibility for their behavior.

To begin with, Aquinas asked why the created order resembled a Great Chain containing a myriad of plants and animals in a descending hierarchy of importance. St. Thomas concluded that the proper workings of nature depended upon the labyrinth of relationships, obligations, and dependencies among God's creatures. Geographer Clarence J. Glacken of the University of California says that as far as St. Thomas was concerned, God had intended that nature be populated with "many creatures differing among themselves in gradation of intellect, in form, and in species."[2] According to Aquinas, "diversity and inequality"[3] guaranteed the orderly working of the system as a whole. The churchman reasoned that if all creatures were equal, they could not "act for the advantage of another."[4] By making each creature different, God established a hierarchy of obligations and mutual dependencies in nature.

St. Thomas's characterization of nature bears a striking likeness to institutional arrangements in medieval Europe, where there existed a tightly defined social structure in which everyone's individual survival depended upon the dutiful performance of a complex set of mutual obligations within a rigidly maintained hierarchical setting. From serf to knight, from knight to lord, and from lord to Pope, all were unequal in degree and kind, each was obligated to the other by the medieval bonds of homage, and all together made up a mirror in which could be viewed, though only hazily, the perfection represented in God's total Creation. According to the late historian Robert Hoyt of the University of Minnesota:

> The basic idea that the created universe was a hierarchy, in which all created beings were assigned a proper rank and station, was congenial with the feudal notion of status within the feudal hierarchy, where every member had his proper rank with its attendant rights and duties.[5]

As people's relationship to their environment has changed, so too have their concepts of nature. Each resulting cosmology has borne the unique imprint of the special circumstances that confronted the human family at a given time and place in history. All cosmologies, however, share the same overarching theme. They tend to serve as a distant mirror of the day-to-day activity of a civilization.

This is not to suggest that people's cosmologies are mere fabrications, as many social relativists claim. Some social critics would have us believe that our cosmologies have no real footing at all in the external world. The social relativists contend that our ideas about nature are completely subjective and bear no resemblance to the world as it exists in fact. While they are right in assuming that our ideas about nature are socially biased and deeply influenced by the cultural context in which we live, they are wrong in assuming that such ideas are without a basis in the "real" world. The fact is, our cosmologies are based on the workings of the real world, but only that small portion of the real world where society and nature interact. People learn things about nature in the process of organizing it. The things that they learn are useful. They allow people to interact with nature, to manipulate and appropriate it. The problem is that people take the things that they have learned about nature and puff them up in such a way as to create an all-encompassing explanation of the workings of the cosmos. Cosmologies, then, are distortions. They are society's way of inflating its rather limited "real world" relationship to the environment, at any given time, into universal truth. Cosmologies are made up of small snippets of physical reality that have been remodeled by society into vast cosmic deceptions.

## Darwin's Nature and the Industrial Mind

Charles Darwin's theory of evolution has proven to be a very compatible companion to the Industrial Age. It is no secret that Darwin's theory of evolution has been exploited over and over again to justify various political and economic ideologies and interests. Social Darwinism has been examined, debated, and analyzed for over a hundred years. In virtually all the discussions of Social Darwinism there is an underlying assumption that the theory itself is a disinterested, objective, impartial recording of nature's operating design, untainted by social context and cultural bias. It is assumed that what Darwin discovered is a law of nature and that society then exploited it for political ends. A new generation of scholars, however, is beginning to question the theory itself, suggesting that in its very conception it might have been as socially biased as the ends to which it was later used.

It should be noted that the scientific and social critiques of Darwin's theory are not critiques of evolution itself, as it is widely accepted, in virtually every academic quarter, that life on Earth has indeed evolved. Rather, a growing number of critics, both in the life sciences and the social sciences, are questioning Darwin's particular version of how the evolutionary process has unfolded, arguing that his views were heavily influenced by the prevailing social gestalt of the times.

Otto Rank suggests that Darwin's theory was just the English bourgeoisie looking into the mirror of nature and seeing their own behavior reflected there.[6] Although such comments are unlikely to grace the pages of any introductory book on biology, it remains a fact that Darwin was a product of his time and subject to the flights and fancies that embroidered the Victorian landscape. It can hardly be a matter of doubt, says University of Connecticut historian John C. Greene, that "like every other scientist, Darwin approached nature, human nature, and society with ideas derived from his culture."[7] That being the case, to understand Darwin's theory of biological evolution it is necessary to understand the economic, social, and political environment that provided the imagery that he used so artfully to sketch his "creation."

Darwin's life spanned the very years that marked the transition from an agrarian economy to the Industrial Age of capitalism. England was at the forefront of the revolutionary changes that were transforming the economic life of Europe. Having a head start on her Continental neighbors, she needed a new cosmology that could make sense of and be compatible with the disorienting array of economic changes that were turning England from a land of haystacks into a land of smokestacks.

Writing in the *Quarterly Review of Biology*, the late Alexander Sandow observed that "Darwinism sprang up where and when capitalism was most strongly established."[8] Historian Greene notes, "British political economy, based on the idea of the survival of the fittest in the marketplace, and the British competitive ethos generally predisposed Britons to think in terms of competitive struggle in theorizing about plants and animals as well as man."[9]

What made Darwin's cosmology so terribly engaging was that it so conveniently fit the age for which it was written, says biographer Geoffrey West.

> In the machine age he established a mechanical conception of organic life. He paralleled the human struggle with a natural struggle. In an acquisitive hereditary society he stated acquisition and inheritance as the primary means of survival.[10]

Darwin dressed up nature with an upper-class English personality, ascribed to nature English motivations and drives, and even provided nature with the English marketplace and the English form of government. Like others who preceded him in history, Darwin borrowed from the popular culture the appropriate metaphors and then transposed them to nature, projecting a new cosmology that was remarkably similar in detail to the day-to-day life to which he was accustomed.

The economic and commercial dealings of the British marketplace provided a number of fruitful metaphors for Darwin's own meanderings on nature. For example, the British naturalist saw the same principle of division of labor at work in nature as in the new industrial factories. Darwin argued:

> [T]he greatest number of organic beings (or more strictly the greatest amount of life) can be supported on any area, by the greatest amount of their diversification . . .[11] For in any country, a far greater number of individuals descended from the same parents can be supported, when greatly modified in many different ways, in habits, constitution and structure, so as to fill as many places, as possible, in the polity of nature, than when not at all or only slightly modified.[12]

By finding in nature the same kind of division of labor at work as that found in the English factory system, Darwin provided a "scientific guarantee of the rightness of the property and work relations of industrial society."[13] Henceforth, capitalist owners could justify the new factory system, with its dehumanizing process of division of labor, by claiming that a similar process was at work in nature. In short, says historian of science Robert Young of Cambridge University in England, Darwin gave "the mark of scientific respectability to the equation of the division of labor with the laws of life."[14]

Similarly, Darwin's concept of divergence in nature provided an ideal defense of English imperialism during the heyday of its colonial expansion. Darwin argued that occasionally a new organism will exhibit new traits sufficiently different from those of its peers to allow it to fill a previously unoccupied niche in nature. Migration into new niches lessens the competition for existing slots and at the same time opens up entirely new areas for exploitation. As Donald Worster points out in *Nature's Economy,* diversity, for Darwin, was "nature's way of getting round the fiercely competitive struggle for limited resources."[15] For the millions of Englishmen forced to leave the British Isles in the nineteenth century to look for new economic opportunities in alien lands, Darwin's notion of divergence made more than a little sense. In the colonies, there were, as yet, untapped opportunities, economic niches ready to be filled and exploited. In contrast to the niggardly supply of economic possibilities that presented themselves on the depressed English homefront in the 1830s and 1840s, divergence seemed a welcome reprieve. Moreover, at a time when Britain was extending her influence into the remote regions of the globe, colonizing new lands and peoples, it was most reassuring to know that wherever the Union Jack was raised, natural selection was being allowed to flourish.

Darwin's ideas about how organisms behave in nature also had much in common with economist Adam Smith's ideas on how buyers and sellers behave in the marketplace. Smith argued that an "invisible hand" regulated supply and demand in the marketplace, allowing each person free reign to maximize his or her own self-interest. Smith acknowledged that such behavior is self-serving, but claimed that the very act of individual selfishness benefits the general well-being of others. Darwin agreed with Smith that in nature as in society, each individual organism is absorbed with maximizing its own self-interest and surviving in the struggle with others over limited resources: "Each individual of each species holds its place by its own struggle and capacity of acquiring nourishment."[16] The problem for Darwin was to try to understand how such individual activity benefitted the entire species. Darwin reasoned that just as an external law—the invisible hand—is constantly at work in the economic sphere, regulating and balancing supply and demand, a similar law—natural selection—must be constantly at work in nature, forever regulating and balancing the supply of resources against the demand for those resources.

Likewise, whereas each individual organism is interested in only its own survival, its triumph can't help but advance the common good, since its traits live on in its offspring, thus assuring a never-ending process of gradual improvement in the biological characteristics of the species as a whole. Philosopher Peter J. Bowler of Queens University in Belfast, Ireland, captures the close correlation between the invisible hand and Darwin's theory of evolution. He observes that "both the balance of nature and the *laissez-faire* view of competition were based on the belief that nature and society are fundamentally harmonious systems in which apparent conflict serves for the benefit of all."[17]

Darwin's description of the evolution of species also relied heavily on machine imagery. There was simply no way to escape the overwhelming presence of the machine in English life at the time. Here was this marvelous new technology that was reshaping the world. "Naturally," everyone was anxious to extend its application to every facet of life. One could hardly expect naturalists to remain aloof from the excitement of the day.

Before the age of the machine, living creatures were viewed as "wholes." This idea fit well with an artisan mode of production in which the craftsman "molded" his creation from a primordial substance. This traditional view of nature was overthrown and replaced with a radically new conception, compatible with the new form of industrial production.

Darwin came to view living things as the sum total of parts "assembled" together into more complex and efficient living machines. Darwin admitted that it was no longer possible for him even to imagine that living creatures were created whole and in their entirety. With the idea of step-by-step machine assembly so firmly implanted in the British mind, Darwin concluded:

> Almost every part of every organic being is so beautifully related to its complex conditions of life that it seems as improbable that any part should have been suddenly produced perfect, as that a complex machine should have been invented by man in a perfect state.[18]

The world historian Oswald Spengler placed the theory of evolution in its most succinct context with the observation that Darwin's entire thesis amounted to little more than "the application of economics to biology."[19] Darwin's ideas attracted widespread support because they seemed to explain the nature of things in terms that were easily recognizable. Bewildered by the changes sweeping over their lives, people were anxious for some kind of grand explanation that could put everything into focus. Darwin's theory met the test, precisely because it was able to find in nature the same forces at work as people were experiencing in their day-to-day lives in the factories and towns that pioneered the Industrial Age. So it did, but if the application to nature was motivated by economics, the transference of nature's laws back to society was motivated by politics. With the publication of *The Origin of Species,* the bourgeois class could rationalize its economic behavior by appealing to the universal laws of nature as its ultimate authority. It was possible, even acceptable, to justify the brutal exploitation of the working poor and imperialist adventures abroad all in the name of faithfully obeying the "laws of nature." Historian Gertrude Himmelfarb surveys the political repercussions of what has come to be known as Social Darwinism and concludes that it has served as the undisputed centerpiece for the politics of the Industrial Age. As a political instrument,

> Darwinism has exalted competition, power and violence over convention, ethics, and religion. Thus, it has become a portmanteau of nationalism,

imperialism, militarism, and dictatorship, of the cult of the hero, the superman, and the master race.[20]

## The Evolution of Information

Now, on the eve of a new century, we are again undergoing a revolutionary transformation in our resource base, our mode of technology, and the way we organize economic and social activity. Not surprisingly, these changes are being accompanied by a revised cosmological narrative. New theories about evolution, steeped in information theory and borrowing heavily from cutting-edge ideas in physics, chemistry, and mathematics, are beginning to exert an increasing influence on the fields of evolutionary and developmental biology. Like Darwin's theory, the new ideas about evolution are already beginning to provide an account of nature's operating design that is remarkably compatible with the operational principles of the new technologies and the emerging new global economy. The laws of nature are being rewritten to conform with our latest manipulation of the natural world, allowing us to rationalize the new technological and economic activity of the Biotech Century as a mere reflection of the "natural order" of things.

The new ideas about evolution make up the final strand in the operational matrix of the Biotech Century and provide the all-important legitimizing context. For that reason, it is essential that the new cosmological narrative be closely examined. Our failure to do so might effectively shut the window to any possible future debate on the particulars of the Biotech Century. That's because, as noted earlier, once the revised ideas about evolution become gospel, debate becomes futile, as people will be convinced that genetic engineering technologies, practices, and products are simply an amplification of nature's own operating principles and therefore both justifiable and inevitable.

Many of the new evolutionary models owe much to the conceptual work of mathematician and philosopher Alfred North Whitehead in the early years of the twentieth century. Whitehead is the father of process philosophy and his ideas helped lay the foundation for the emerging theories of evolutionary development. Whitehead starts with the assumption that all of nature consists of patterns of activity interacting with other patterns of activity. Every organism is a bundle of relationships that somehow maintains itself while interacting with all the other relationships that make up the environment. In interacting with their environment, organisms are continually "taking account" of the many changes going on and continuously changing their own activity to adjust to the cascade of activity around them. This "taking account," according to Whitehead, is the same as "subjective aim," meaning that every organism in some way anticipates the future and then chooses one among a number of possible routes to adjust its own behavior to what it expects to encounter. In other words, every organism exhibits some degree of aim or purpose. If an organism were not able to anticipate the future and adjust its behavior to what is about to come, it could not possibly survive all the abrupt changes in the pattern of activity around it. The late Hans Kalmus, of University College in London, observes that "anticipatory actions occur widely in the organic world. A predator catching a moving prey, a tennis player hitting a ball, a spider constructing a web, even a flower displaying its visual and olfactory attractions, all can be said to anticipate future events in their environments."[21]

This constant "anticipation and response" is the central dynamic of all life. A closer look tells us that "subjective aim" is just another expression for mind. Whitehead sees mind (or subjective aim) as existing at every level of life. Organisms are constantly anticipating the future and making choices on how to respond to it. This is mind operating. Evolution is no longer viewed as a mindless affair, quite the opposite. It is mind enlarging its domain up the chain of species.

Whitehead provided the philosophical vision for a new approach to biological evolution. It was Norbert Wiener, however, who provided the much-needed scientific framework. Cybernetics reduces Whitehead's description of mind in nature to quantifiable proportions, replacing any vitalistic description with a purely technological definition of behavior.

For the cybernetician, information feedback and information processing serve as a kind of all-embracing scientific description of how organisms anticipate and respond to changing conditions over time. The cybernetician views living organisms as in . . . formation. A living organism is no longer seen as a permanent form but rather as a network of activity. With this new definition of life, the philosophy of becoming supersedes the philosophy of being, and life and mind become intricately bound to the notion of "processing" change.

Although cybernetics is largely concerned with how systems maintain themselves over time, the new theory of development also makes room for the idea of evolutionary change in systems by way of positive feedback. Ilya Prigogine, a Belgian physical chemist and Nobel Laureate, has devised a theory of "dissipative structures" to explain how cybernetic principles can incorporate the notion of evolution as well as homeostasis. According to Prigogine, all living things and many nonliving things are dissipative structures. That is, they maintain their structure by the continual flow of energy through their system. That flow of energy keeps the system in a constant state of flux. For the most part, the fluctuations are small and can be easily adjusted to by way of negative feedback. However, occasionally the fluctuations may become so great that the system is unable to adjust and positive feedback takes over. The fluctuations feed off themselves, and the amplification can easily overwhelm the entire system. When that happens, the system either collapses or reorganizes itself. If it is able to reorganize itself, the new dissipative structure will always exhibit a higher order of complexity and integration, and a greater flow-through, than its predecessor. Each successive reordering, because it is more complex than the one proceeding it, is even more vulnerable to fluctuations and reordering. Thus, increased complexity creates the condition for evolutionary development.

Like Prigogine, many scientists are coming to view evolution as the steady advance toward "increased complexity of organization."[22] Organizational complexity, in turn, "is equivalent to the accumulation of information."[23] In other words, evolution is seen as improvement in information processing. The more successful a species is at processing more complex, more diverse kinds of information, the better able it is to adjust to a greater array of environmental changes. By this new way of thinking, the key to evolution itself is to be found in how information is processed. Negative feedback leads to stasis. Positive feedback leads to transformation.

Much of the impetus for the new ideas about evolution are coming from mathematicians, physicists and biologists working in the new "sciences of complexity." While many divergent strains of thought are emerging in the field of complexity theory, there is a general consensus that the

Darwinian emphasis on natural selection as the prime mover of evolution, although important, is not sufficient to explain the origin and development of species. Biologists like Brian Goodwin, of the Open University in England, argue that organisms are more than the sum of their DNA. Goodwin says that just as important in development is "the relational order among components . . . the way they are organized in space and how they interact with one another in time." In complex systems, notes Goodwin, "chaotic behavior at one level of activity—molecules or cells or organisms—can give rise to distinctive order at the next level—morphology and behavior." In other words, "order emerges out of chaos."[24]

In this revised way of thinking about evolution, organisms are no longer viewed as passive beings resulting from the random process of natural selection, but rather, as dynamical self-organizing processes in which the organisms continually create themselves by ordering their activity into coherent emergent wholes. "Organisms cease to be mere survival machines" and come to look more like "works of art."[25] Murray Gell-Mann, a Nobel Prize–winning scientist from the California Institute of Technology, notes that organisms are complex adaptive systems that are "pattern seekers," that is, they continually interact with their environment, learn from their experience, and adapt. The evolutionary process, then, is creative as well as random, self-organizing as well as selective. Evolution is viewed in Whiteheadian terms as "the creative advance into novelty."[26]

Norman Packard, formerly of the University of California at Santa Cruz, and Chris Langton of the Santa Fe Institute use the term "life at the edge of chaos" to refer to the region where "all parts of the [living] system are in dynamic communication with all other parts, so that the potential for information processing in the system is maximal." "It is this state of high communication and 'emergent' computation . . . that provides maximal opportunities for the system to evolve dynamic strategies of survival," says Goodwin.[27] "Survival," in the new way of thinking, has to do with gathering information about the environment, and responding appropriately. In complex dynamical systems "what drives their evolution is increased computational ability," says Packard.[28]

The new view of evolution as increased computational ability is gaining supporters within the biological community. Biologist Edward O. Wilson of Harvard University observes, "There's been a general increase in information processing over the last 550 million years, and particularly in the last 150 million years."[29] While many mainstream evolutionary biologists continue to be uncomfortable with the idea that evolution represents an increase in computational complexity, others have been won over to the idea. Francisco Ayala, professor of ecology and evolutionary biology at the University of California at Irvine, says, "The ability to obtain and process information about the environment, and to react accordingly, is an important adaptation because it allows the organism to seek out suitable environments and resources and to avoid unsuitable ones."[30] Norman Packard sums up much of the new thinking about evolutionary change:

> It seems reasonable that the task of survival requires computation. If that's true, then selection among organisms will lead to an increase in computational abilities. That creates an arrow of change, not just a drift upward.[31]

The story of Creation is being retold. This time around, nature is cast in the image of the computer and the language of physics, chemistry, mathematics, and the information sciences. With living organisms, as with computers, information capacity and time constraints become the primary considerations. Each succeeding species up the evolutionary chain, like each new generation of computers, is more complex and better adept at processing increasing amounts of information in shorter periods of time.

Much of the new thinking about evolution parallels the new way commerce is being organized in the network-based global economy. The traditional industrial market, based on individual firms competing with each other in a Darwinian environment of "survival of the fittest," while still the dominant mode, is beginning to give way to new forms of commerce based on the creation of shared relationships within complex embedded networks. Being able to anticipate and respond quickly to fast-changing commercial environments is the key to survival and growth in the new era. Success, therefore, is increasingly measured by the ability to manage a growing diversity of information, which requires more complex operational webs. Companies are finding that the joining together of diverse and often competing and conflicting business interests often creates wholly new emergent forms of business with new properties and possibilities that are qualitatively different than the sum of the practices of the individual firms—a clear example of "creative advance into novelty" as a result of reorganizing relationships at "the edge of chaos." In the coming century, more and more commerce, especially in cyberspace, but also in geographical markets, will be embedded in increasingly complex computational networks whose webs will span the globe.

In a society of increasing complexity, in which the collecting, exchanging, rearranging, and discarding of information is proliferating at an unparalleled speed, and in which personal and institutional success is measured in terms of the ability to process increasingly complex amounts of information, it is easy to see why biologists might come to see the same forces at work in nature.

It's also not hard to understand why the first generation raised in a fully computerized society might come to accept so readily the new concept of nature that is emerging. They are growing up using the computer to organize their entire environment. Is it any wonder, then, that they will come to believe that nature itself is organized by the same set of assumptions and procedures they themselves are using when they manipulate it? The point is, our new ideas about how nature operates are coming to mirror the new technological and economic relationships we are establishing with the natural world, providing a new generation the assurance that the way it's going about organizing its world is compatible with nature's own organizing design.

For example, humanity is coming to view nature in "computational" terms at the same time as scientists are using sophisticated genetic techniques to program the future performance of living things. It is now possible to program a new gene into an organism before birth in order to anticipate and effect a change in the activity of that organism years later. Being able to anticipate and respond to the future state of an organism and its environment by programming changes in its genetic information at conception represents the ultimate use of cybernetic principles and information processing skills. Now that we are using information processing skills to program the future of living "systems," we are also coming to see information processing and increasing computational ability as a distinguishing feature of the evolutionary schema in nature.

This is only the beginning of a long string of convenient "coincidences." For example, now that we are continually reengineering the genetic instructions of organisms to make them compatible with the fast-changing artificial environments we have created, we contend that organisms in nature are continually reengineering themselves as they develop in order to adjust to their own fast-changing environments. In the revised approach to evolutionary development, "self-organization," by way of negative and positive feedback, becomes as important as "natural selection" in securing the survival of both the organism and its offspring. Every organism seeks to maximize its own self-organization by exchanging information with its environment and anticipating and responding to a range of novel events. While each species seeks only its own self-organization, in the process it generates new bits of information, which are the source of further evolutionary development. Every new evolutionary advance, in turn, increases the overall complexity of the system, further integrating all the information into a richer labyrinth of relationships. In an age steeped in the information mystique, people can take comfort in the belief that their own efforts to generate and exchange greater amounts of information not only advance their self-organization but, as is the case with natural evolution, also contribute to the strengthening of all societal relationships by increasing the level of interaction, interconnection, and complexity in the system as a whole.

Similarly, now that we are able to change an organism's characteristics quickly with the new gene technologies, we have come to believe that fundamental evolutionary changes in nature occur quickly as well. A new generation of scientists is advancing the idea of "punctuated equilibrium," which argues that basic biological changes in nature happen suddenly and rapidly, not slowly and piecemeal, as Darwinists have long contended. It is also interesting to note that according to the punctuated-equilibrium theory new species often develop in total isolation from the parent stock. Of course, in the biotechnical age such will also be the case. Scientists will be able to create new transgenic organisms by manipulating genetic material in a totally isolated laboratory environment and then release the transgenic creatures into the Earth's many ecosystems.

Cosmologies serve still another rationalizing function in every society. Every time the human family changes the way it goes about expropriating the natural world, it finds it necessary to sever any sense of empathetic association it might feel toward the objects of its assimilation. It's much more difficult to exploit something you identify with, so concepts of nature serve as a kind of psychic ritual by which human beings deaden the natural world, preparing it for consumption.

Darwin's world was populated by machine-like creatures. Nature was conceived as an aggregate of interchangeable parts assembled into various functional combinations. This mechanical conception of living beings robbed sentient creatures of any remaining sacred qualities. The denaturing and mechanizing of the biological kingdom eliminated intrinsic value and replaced it with John Locke's notion of utility value. Most scientists and much of the public came to share René Descartes' view of living creatures as "soulless automata," whose movements were little different from those of the automated puppetry that danced upon the Strasbourg clock.

The revised notions of evolution replace the idea of life as machinery with the idea of life as information. By resolving structure into function and reducing function to information flows, the new cosmology all but eliminates the idea of species integrity. Living things are no longer perceived as birds and bees, foxes and hens, but as bundles of genetic information. All living beings are drained of their substance and turned into abstract messages. Life becomes a code to be deciphered. There is no longer any question of sacredness or specialness. How could there be when there are no longer any recognizable boundaries to respect? In the new way of thinking about evolution, structure is abandoned. Nothing exists in the moment. Everything is pure activity, pure process. How can any living thing be deemed sacred when it is just a pattern of information?

Eliminating structural boundaries and reducing all living entities to information provides the proper degree of desacralization for the bioengineering of life. After all, in order to justify the engineering of living material across biological boundaries, it is first necessary to challenge the whole idea of an organism as an identifiable, discrete being, with a permanent set of attributes. In the age of biotechnology, separate species with separate names gradually give way to systems of information that can be reprogrammed into an infinite number of biological combinations. It is much easier for the human mind to accept the idea of engineering a system of information than it is for it to accept the idea of engineering a dog, chimpanzee, or human being. In the coming age it will be much more accurate to describe a living being as a very specific pattern of information unfolding over a period of time.

In the new scheme of things, each species is "better informed" than its predecessor and thus better equipped to anticipate and control its future. If evolution is the increase in computational ability, then humanity is performing its proper role in the cosmic scheme by its relentless drive to process increasing stores of information in order to anticipate and control its own future. Kenneth M. Sayre, a professor of philosophy at the University of Notre Dame, accurately describes humanity's newest rationale for the manipulation of nature when he writes:

> Human beings . . . excel in the acquisition of information, and also in versatility of information processing . . . Since superiority in information gathering and processing amounts to superior adaptive capacities, this accounts for human dominance over other kinds.[32]

Suddenly the old Darwinian notion of "survival of the fittest" is replaced by the idea of "survival of the best informed." Mental acumen, not brute force, becomes the key to evolutionary advancement. Human beings, the best "information processors" in the biological kingdom, are now advancing the evolutionary process by downloading genetic information and reprogramming nature using engineering design principles and genetic engineering tools.

The new operations approach to the engineering of life as well as the cosmological justification for going ahead with it was first advanced more than half a century ago by Norbert Wiener: "It is my thesis that the physical functioning of the living individual and the operation of some of the newer communication machines are precisely parallel in their analogous attempts to control entropy through feedback."[33] That being the case, there is no reason whatsoever why bioengineering shouldn't proceed. If, as Wiener and his proteges in the fields of engineering and biology contend, living organisms and machines closely resemble each other, then bioengineering is just an amplification of nature's own operational principles. As such, bioengineering is merely a logical extension of, but hardly a radical departure from, the way nature itself operates.

The new language of evolution seems to suggest that nature has always operated much the same way we are operating when we engineer it in the laboratory. Of course, in a sense, the new cosmology contains a germ of truth. If nature didn't exhibit some of the characteristics we ascribe to it, we would find it impossible to manipulate it the way we do in the laboratory. The problem, once again, is that in our cosmologies we inflate the tiny aspect of nature's reality that we are manipulating at a moment in time into a universal cosmology and then claim that all of nature operates in a manner that is congenial with the way we are operating. We continually remake nature to suit our own needs and then conclude that the technological procedures we are using at the time must be similar to the procedures used in constructing the original creation.

## A Postmodern Cosmology

Every society is an organizational expression of humanity's deep desire to overcome the limits imposed by time and space. The goal is always the same. We organize to perpetuate ourselves, and our dream is to organize ourselves so well that we will be able to overcome our temporal sojourn and experience some measure of earthly immortality. In the Middle Ages, skilled craftsmen and chemists dreamed of turning lead into gold and of mixing chemicals with fire to discover the secret elixir that would guarantee everlasting life. With the dawn of the Industrial Age, people's dreams turned from alchemy to perpetual motion. Generations of inventors and machinists gave over their lives and their fortunes in their quest to build the perfect machine; one that would run by itself, be totally self-contained, and thus live on forever.

The age of biotechnology incorporates its own unique vision of immortality, best expressed in a popular television series seen by millions of people throughout the world. In *Star Trek,* the starship *Enterprise* contains a special room called the transporter room. Personnel wishing to leave the starship do so by entering the transporter room. The transporter itself is a sophisticated computer that acts as a "matter/energy scrambler."[34] According to Captain Kirk, the transporter "converts" matter temporarily into energy, beaming that energy to a fixed point. In other words, the human body is transformed into billions of bits of information, which are then sent through space by way of electronic pulses. The information is then downloaded and reassembled at its destination, restoring the body to its original form. The ship's doctor, McCoy, points out that people can even be "suspended in transit until a decision is reached to rematerialize them."[35] The transporter ". . . retains the memory of the original molecular structure of everyone and everything passing through it,"[36] allowing it to turn information from matter to energy and back to matter again.

The transporter room is clearly science fiction. Still, it speaks to an eternal yearning that is as seductive to some molecular biologists as perpetual motion was to industrial engineers and alchemy to medieval metallurgists. The ability to reduce all biological organisms and ecosystems to information and then to use that information to overcome the limitations of time and space is the ultimate dream of biotechnology. Norbert Wiener understood as much and anticipated the idea of a transporter room more than a half century ago. Wiener starts off with the assumption that all living things are really "patterns that perpetuate themselves," and that "a pattern is a message and may be transmitted as a message."[37] Wiener then asks:

> What would happen if we were to transmit the whole pattern of the human body . . . so that a hypothetical receiving instrument could re-embody these messages in appropriate matter, capable of continuing the processes already in the body and the mind . . . ?[38]

Wiener concludes that "the fact that we cannot telegraph the pattern of a man from one place to another seems to be due to technical difficulties," but he hastens to add that "the idea itself is highly plausible."[39] It is likely that the genetic engineers will never completely work out those "technical difficulties," but that's of little matter. The alchemists of the Medieval Age and the engineers of the Industrial Age were never able to work out the "technical difficulties" of creating the elixir or producing perpetual motion; still, their visions provided a distant goal for their journey.

Many scientists on the cutting edge of the computer technology revolution are convinced that information is the key to immortality. Yoneji Masuda, a leading figure in the Japanese plan to become the first fully developed information society, writes enthusiastically of this newest vehicle to earthly immortality.

> Unlike material goods, information does not disappear by being consumed, and even more important, the value of information can be am-

> plified indefinitely by constant additions of new information to the existing information. People will thus continue to utilize information which they and others have created, even after it has been used.[40]

Information, it is argued, is impervious to the ravages of time. It is of this world but does not die with the flesh. Just as a Christian might contend that the body is merely a temporary vessel for the everlasting spirit that resides in it, the new cosmologists would contend that the body is merely a temporary vessel for the information that is embedded in it. Gerald Jay Sussman, Massachusetts Institute of Technology professor of electrical engineering and computer science, expresses the hopes and expectations of many of his colleagues:

> If you can make a machine that contains the contents of your mind, then the machine is you. To hell with the rest of your physical body, it's not very interesting. Now, the machine can last forever. Even if it doesn't last forever, you can always dump it onto a tape and make backups, then load it up on some other machine if the first one breaks. . . . Everyone would like to be immortal. . . . I'm afraid, unfortunately, that I am the last generation to die.[41]

Similarly, many molecular biologists see the information contained in DNA as immortal. Some researchers are storing the germplasm of extinct plants and animals in special gene banks. They hope to reconstruct the living forms sometime in the future by learning how to set in motion the information contained in the genetic codes.

While the prospect of bringing extinct creatures back to life is speculative, human cloning is a very real possibility in the next decade. With human cloning, one's genetic information can be replicated endlessly into the future, creating a kind of pseudo-immortality. It's not so difficult to imagine one wanting his or her genetic likeness to live on in the form of a cloned offspring, ensuring at least a modicum of earthly immortality. The desire for immortality, if not in this world, then at least in the next, has inspired the creation of all of the world's major religions and been the driving force behind every great civilization in history. Why should passing along one's genetic inheritance *in toto* be looked on any less favorably than passing on one's material inheritance? What could be more personal than the "property" one has in his or her own genetic makeup? In a society that places such high regard on the rights of property owners, why shouldn't every person be allowed to perpetuate his or her most cherished possession—their genotype—by means of clonal propagation?

What about cloning others? A grieving spouse or parent might want to clone a copy of a dying partner or child with the hope of having their likeness "live on." Some scientists have suggested that "superior" individuals might be maintained endlessly through clonal propagation. Dr. Joshua Lederberg has written,

> If a superior individual—and presumably, then, genotype—is identified, why not copy it directly, rather than suffer all of the risks, including those of sex determination, involved in the disruptions of recombinations (sexual procreation). Leave sexual reproduction for experimental purposes: when a suitable type is ascertained take care to maintain it by clonal propagation.[42]

Succeeding generations may well seek after the genomic information embedded in living processes the way past generations sought after the elixir and perpetual motion. Storing and perpetuating complex bundles of genetic data could become the immortality image of the coming age.

In the new cosmology, then, it is information that is elevated and made the primary object of our attention. According to the new thinking, the evolution of information is tantamount to the evolution of life. With information's new, exalted status, even knowledge itself is losing its meaning and increasingly being subsumed under the category of information. Today, "to be knowledgeable" and "to be informed" have come to mean virtually the same thing. This is a revolution in the history of consciousness. By changing the meaning of knowledge, so that "to know" becomes equivalent to "being informed," we saturate knowledge with temporality. "To be informed" means to be aware of changing conditions. Being informed requires a constant updating. It is an ongoing process of anticipation and accommodation to changes going on in the environment. To be knowledgeable today is to be continually aware of the changes going on around us and to be able to adjust accordingly. Knowledge, in the new scheme of things, is no longer viewed as discovery of facts but rather as an ongoing creative process.

In all earlier periods of history, knowledge came slowly. For that reason, each new insight was enshrined and closely guarded. Knowledge lasted because great lapses of time occurred between insights. That's no longer the case. What we know today is quickly eclipsed by what we will know tomorrow. Thus, we have far less toleration for timeless truths and ironclad laws. By their very nature, timeless truths and ironclad laws impose boundaries and establish limits. They tell us what we can and cannot do. In every cosmology, they serve notice as to how far it is possible to go.

In our nanosecond culture everything changes so fast that it is necessary to construct a cosmology in which change itself is honored as the only timeless truth. By reinterpreting nature as the evolution of information, humanity achieves this end. Nature is no longer seen as a set of restraints, but rather as a process of "creative advance." Knowledge gathering is now information gathering, and information gathering is the processing of changing conditions over time. With this linguistic metamorphosis, the door is opened to a thorough reinterpretation of existence as pure process devoid of any kind of ultimate, unchanging frame of reference.

St. Thomas Aquinas and Charles Darwin were both trying to express, as best they could, the workings of nature. They truly believed that their formulations were an act of discovery, an unmasking of the universal scheme of things. They sought the truth and believed that it existed somewhere outside themselves. They were convinced that their cosmologies were an accurate description of the way the world is.

Although scientists continue to search for the underlying truths of reality, hoping to unmask the universal rules and principles that give order and meaning to existence, their notion of a knowable, objective reality is being increasingly challenged by a new generation of "postmodern" scholars who take exception to the very idea of universal truths and grand cosmological stories. They argue that there are no overriding meta-narratives or universal truths, but only playful options and culturally constructed and socially construed scripts and myths. Life is viewed less as a journey of discovery and more as a creative adventure.

The postmodernists point to the emergence of a whole new vocabulary of words and terms as proof, of sorts, that the self-deception that has guided our cosmologies over the millennia is about to be expurgated once and for all. They suggest that the very idea of an "objective" reality is giving way to the idea of "perspective" realities. The idea that future states are subject to ironclad laws of causality is giving way to the idea that the future is a trajectory of "creative possibilities." The idea of "deterministic outcomes" is being replaced with the idea of "likely scenarios." The idea of "permanent truths" is being replaced with the idea of "useful models." Many postmodernist scholars are convinced that this change in vocabulary will free us from the long-standing hubris of believing we have discovered the absolute truths about nature.

Ironically, the new language, which many postmodern thinkers interpret as an expression of deep humility, may, in fact, be a more subtle expression of an invasive cultural narcissism, laying the philosophical groundwork for the Biotech Century and a Brave New World. Far from being a repudiation of humankind's long history of cosmological projections, the postmodernists' views may be simply the newest version—one that conforms more closely to our new technological and economic relationship to the natural world.

"Perspectives," "scenarios," "models," "creative possibilities." These are the words of a creator, an architect, a designer. Perhaps humanity is abandoning the idea that the universe operates by ironclad truths because it no longer feels the need to be constrained by such fetters. Nature is being made anew, this time by human beings. We no longer see ourselves as guests in someone else's home and therefore obliged to make our behavior conform with a set of preexisting cosmic rules. It is our creation now. We make the rules. We establish the parameters of reality. We create the world, and because we do, we may no longer feel beholden to outside forces and universal truths. We may no longer have to justify our behavior, now that we are taking on a new task as architects of our own biological destiny and the rest of nature. Viewing nature as a "creative advance into novelty" and each species as "a work of art" serves the ends of a eugenics future. If nature as a whole is an evolving work of art, then our species is justifiably the ultimate artist, whose evolutionary mission is to continually shape and mold our own nature and the rest of nature to reflect our own artistic sensibilities.

The new ideas about evolution offer much more than a convenient rationale. They delineate humanity's new responsibility. One hundred years after Thomas Huxley's eloquent defense of Darwin's theory, his grandson, the late Julian Huxley, took up the banner of the emerging cosmology. Humanity, itself the product of evolutionary creativity, is now obligated, said Huxley, to continue "the creative process" by becoming the architect for the future development of life. *Homo sapiens'* destiny, he said, is to be "the sole agent of further evolutionary advance on the planet."[43] Humanity's new responsibility is momentous. All Darwin asked of people was that they compete for their own life. The new cosmology asks people to be "the creator" of life. Huxley believed that we have no other choice but to become the "business manager for the cosmic process of evolution."[44]

Genetic engineering, Huxley and other biologists would argue, is the inevitable result of Whitehead's "creative advance into novelty" that began with the emergence of the first organism. From the very beginnings of life, each organism has striven to enlarge its informational domain, to become "better informed." That the human mind has now become so well informed that it could actually conceive of using the vast amount of information at its disposal to engineer life is itself an acknowledgment of the entire evolutionary process at work.

Therefore, if one accepts the new explanations for how life organizes itself, one has little choice but to accept bioengineering as well. Not to do so would appear to violate the very process of evolutionary development. By the new cosmological thinking, bioengineering is not something artificially superimposed on nature but something spawned by nature's own ongoing evolutionary process. It is, in effect, the next stage in the evolutionary process. Any effort, therefore, to resist bioengineering would in the end be futile and self-defeating because it would fly in the face of what is "natural."

A growing number of scientists already view themselves as the architects of a new evolutionary course for the human race. In his book *Remaking Eden,* molecular biologist Lee Silver muses about a distant future for which he and his colleagues in the genetic sciences are blazing the trail.

> In this [new] era, there exists a special group of mental beings. Although these beings can trace their ancestry back directly to *Homo sapiens,* they are as different from humans as humans are from the primitive worms with tiny brains that first crawled along the Earth's surface. It is difficult to find the words to describe the enhanced attributes of these special

> people. "Intelligence" does not do justice to their cognitive abilities. "Knowledge" does not explain the depth of their understanding of both the universe and their own consciousness. "Power" is not strong enough to describe the control they have over technologies that can be used to shape the universe in which they live.
>
> These beings have dedicated their long lives to answering three deceptively simple questions that have been asked in every self-conscious generation of the past. "Where did the universe come from?" "Why is there something rather than nothing?" "What is the meaning of conscious existence?"
>
> Now, as the answers are upon them, they find themselves coming face-to-face with their creator. Whom do they see? Is it something that twentieth-century humans can't possibly fathom in their wildest imagination? Or is it simply their own image in the mirror as they reflect themselves back to the beginning of time . . . ?[45]

Today, our biotechnical arts merely imitate nature. Tomorrow, they could subsume it. Our children may be convinced that their creations are of a far superior nature to those from which they were copied. They may come to view their imitation of nature as nature and their art could become their reality.

More than thirty years ago, Dr. Joshua Lederberg wrote expectantly of the possibility of designing "a useful protein from first premises, replacing evolution by art."[46] Recombinant DNA techniques are the "artists' tools" of the postmodern era. With the new technologies, human beings assume the role of creative artists, continually transforming evolution into works of art. This new kind of art, however, is very different from the kind of artistic sensibilities we've known in the past. It is, in a sense, a counterfeit art, steeped in the techniques of rational calculation, mass production, and customization.

Genetic engineering—as an "art form"—epitomizes the new postmodern way of thinking that has grabbed hold of the culture, effecting a broad change in the way we perceive our very being. The new postmodern world in art and architecture, film, television, popular music, and in the increasingly virtual worlds we delight in and travel through, is one of ever fewer boundaries; a place where past, present, and future twist and meld, where life is less serious and more playful and where the rules of engagement are forever changing. The new era is less constrained by fate and destiny and more open to a therapeutic frame of mind in which each person is free to create and live out as many fantasies, experiences, and lifestyles as time permits.

The postmodernists tell us that we are experiencing a shift from an "industrious" age to a "creative" age, a new period in history characterized less by productive output and more by an exuberant artfulness. It is within this context that the new biotechnologies take on such importance. In their near limitless possibilities to reconstruct and reinvent the body, move DNA across species boundaries, erase the genetic past, and pre-program the genetic future, they bring the biology of life squarely in line with the new protean spirit. Life, long thought of as God's handiwork, more recently viewed as a random process guided by the "invisible hand" of natural selection, is now being reimagined as an artistic medium with untold possibilities. Freeman Dyson writes:

> It is impossible to set any limit to the variety of physical forms that life may assume . . . It is conceivable that in another $10^{10}$ years, life could evolve away from flesh and blood and become embodied in an interstellar black cloud . . . or in a sentient computer.[47]

A growing number of young people already see themselves—their very corporeal being—as the ultimate work of art, a continually metamorphosing "project," taking on new shapes, forms and attributes in a never-ending search for new means of self-expression. The widespread popularity of cosmetic surgery, psychotropic mood enhancement drugs, and personal therapies of all kinds are a reflection of the new sense of self as an unfinished work of art. But here the term "art" is grossly misappropriated. To understand why, we need to better distinguish between art and technics.

Human beings have always been image makers and tool makers. We create symbols to communicate our beingness in the world and we fashion technologies to perpetuate our well-being. We are both artists and technicians and the two aspects of our lives have always enjoyed a symbiotic if, at times, ambiguous relationship. The new technologies, however, are blurring the distinction between art and technics. Computer software is, by its very nature, a symbol-making tool. Genetic "wetware," however, is also being redefined as "symbols." In their book *The DNA Mystique: The Gene as a Cultural Icon,* Dorothy Nelkin and M. Susan Lindee note that the gene is fast becoming a reference point in popular culture. In the media, the arts, academia, and in popular lore there is increasing discussion of "selfish genes, pleasure seeking genes, violence genes, celebrity genes, gay genes, couch potato genes, depression genes, genes for genius, genes for saving, and even genes for sinning."[48] The gene, says Nelkin and Lindee, is becoming "a cultural icon, a symbol, almost a magical force."[49] In the popular culture, the biology of the gene is being quickly subsumed by the sociology of the gene. The gene as a producer of images is taking on a social and political role every bit as important as its role as a producer of proteins. Nelkin and Lindee sum up the gene's new status as cultural icon:

> The gene is . . . a metaphor, a convenient way to define personhood, identity, and relationships in socially meaningful ways. The gene is used, of course, to explain health and disease. But it is also a way to talk about guilt and responsibility, power and privilege, intellectual or emotional status. It has become a supergene, used to judge the morality or rightness of social systems and to explore the forces that will shape the human future.[50]

Together, computer software and genetic wetware represent the ultimate "image-making tools," allowing us to use the most sophisticated technics to fashion life into "works of art." It is perhaps understandable that we might prefer to think of the new technologies as artists' tools rather than engineering tools, and ourselves as works of art in process rather than machines being fine-tuned. Altering genetic codes seems more intimate and noble, less cold and inhuman, if it's thought of as an artistic exercise.

By masquerading as artistic tools, the computer and genetic engineering technologies create the illusion that the new era somehow represents a creative renaissance of sorts, a reemergence of the artistic side of the human experience. Rather, the new technologies threaten to smother the artistic sensibility altogether. True art always represents a "deep communion" with the outside world, a reaching out to share with others one's innermost feelings and emotions about the reality we experience. The arts are, at one and the same time, both our most intimate and abstract forms of communication. The intent of the arts is to create a shared space in which the artist can engage others symbolically. Art, Lewis Mumford reminds us, "is essentially an expression of love, in all of its many forms . . . in contrast to technics, which is mainly concerned with the enlargement of human power."[51]

Genetic engineering is, perhaps, the ultimate "enlargement of human power" over life and easily the most advanced form of technics ever conceived. In the sense of the true meaning of art then, genetic engineering is a rank deception. For example, consider the current vogue over "body makeovers." Many in the postmodern culture industry view a body makeover as a vehicle of artistic expression. Others, however, view body makeovers more as a conformist act than a creative act, designed to reengineer a self that will gain the acceptance, approval, and adulation of others. The purpose is often to correct faults and perfect performance, with the goal of achieving some socially mediated norm of acceptability. It is not a projection of one's unique inner being onto the world—a celebration of communion—but, rather, an attempt to remake oneself to "fit in" to the world. It is technique substituting for art.

Making decisions over what genes to insert, recombine, or delete in an effort to "alter," "transform," and "redesign" oneself and one's progeny is less an artistic expression and more a technological prescription. It is not art, but artifice. What some social theorists call the "Creative Age" is really the age of unlimited consumer choice. Unfortunately, we increasingly confuse the ability to choose with the ability to create, especially with regard to the new biotechnologies. Now that we can begin reengineering ourselves, we mistakenly think of the new technological manipulation as a creative act, when in reality it is merely a set of choices purchased in the marketplace. The biotech revolution is, after all, the ultimate consumer playground, offering us the freedom to recast our own biological endowment and the rest of nature to suit whatever whim might move us. More importantly, the new genetic technologies grant us a godlike power to select the biological futures and features of the many beings who come after us—the greatest shopping experience of all time.

# Eight A Personal Note

Over the past twenty years I've expressed growing concern about many aspects of the emerging biotech revolution, leading many in the scientific community, and in the general public, to ask if, in fact, I'm simply opposed to science and the introduction of new technologies. The question is not whether one is in favor or opposed to science and technology writ large, but rather, what kind of science and technology does one favor. I am reminded of the denunciation of critics by the Vatican at the dawn of the modern era. Any views that appeared to challenge the official Church orthodoxy were branded as ungodly and blasphemous, the clear message being there is only one way to believe in God.

The fact is, just as there are many ways to celebrate God, so too are there many ways to celebrate science. We have become so accustomed to thinking of science in strictly Baconian terms that we have lost sight of other approaches to harnessing the secrets of nature. As mentioned in chapter five, Bacon viewed nature as a "common harlot" and urged future generations to "tame," "squeeze," "mould," and "shape" her so that "man" could become her master and the undisputed sovereign of the physical world. Many of today's best-known molecular biologists are heirs to the Baconian tradition. They see the world in reductionist terms, and view their task as grand engineers, continually editing, recombining, and reprogramming the genetic components of life, to create more compliant, efficient, and useful organisms that can be put to the service of humankind. In their research, they often favor isolation over integration, detachment over engagement, and the exercise of applied force or penetration over stewardship and nurturance.

Others in the field of biology, although equally rigorous, exercise a more integrative, systemic approach to nature. The ecological sciences, which are gaining in stature and importance, view nature as a seamless web made of myriad symbiotic relationships and mutual dependencies, all embedded in larger biotic communities that together make up a single living organism—the biosphere. Ecologists favor more subtle forms of manipulation designed to enhance rather than overpower and sever existing relationships, always with an eye toward preserving ecological diversity and maintaining community bonds.

Each of these approaches to the biological sciences lead to very different kinds of practices. For example, in agriculture, we noted in Chapter One that molecular biologists are experimenting with new ways to insert genes into the biological code of food crops to make them more nutritious and more resistant to herbicides, pests, bacteria, and fungi. Their goal is to create a self-contained, safe haven, fortressed away from the large biotic community. Many ecological scientists, on the other hand, are using the new flow of genomic data to better understand the relationship between environmental influences and genetic mutations to advance the science of ecologically based agriculture. Their goal is to combine the wealth of new genetic information being collected with the knowledge being gained on how ecosystems function to establish a more integrative approach to agriculture—one that relies on integrated pest management, crop rotation, organic fertilization, and other sustainable methods designed to make agricultural production compatible with the ecosystem dynamics of the regions where the crops are being grown.

Similarly, in medicine, we noted in Chapters One and Four that molecular biologists are fixing their attention on somatic gene surgery, pumping altered genes into the patient to "correct" disorders and arrest the progress of disease. Their efforts are designed to cure people who have become ill. Other researchers, however, including a small but growing number of molecular biologists, are exploring the relationship between genetic mutations and environmental triggers with the hope of fashioning a more sophisticated, scientifically based understanding and approach to preventive health. More than 70 percent of all deaths in the United States and other industrialized countries are attributable to what physicians refer to as "diseases of affluence." Heart attacks, strokes, breast, colon and prostate cancer, and diabetes are among the most common diseases of affluence. While each individual has varying genetic susceptibilities to these diseases, environmental factors, including diet and lifestyle, are major contributing elements that can trigger genetic mutations. Heavy cigarette smoking, high levels of alcohol consumption, diets rich in animal fats, the use of pesticides and other poisonous chemicals, contaminated water and food, polluted air, and sedentary living habits have been shown, in study after study, to cause genetic mutations and lead to the onset of many of these high-profile diseases.

The Human Genome Project is providing researchers with vital new information on recessive gene traits and genetic predispositions for a range of illnesses. Still, little research has been done, to date, on how genetic predispositions interact with toxic materials in the environment, the metabolizing of different foods, and lifestyle to effect genetic mutations and phenotypical expression. The new holistic approach to human medicine views the individual genome as part of an embedded organismic structure, continually interacting with and being affected by the environment in which it unfolds. The effort is geared toward using increasingly sophisticated genetic and environmental information to prevent genetic mutations from occurring. (It needs to be noted, however, that a number of genetic diseases appear to be unpreventable and immune to environmental mediation.) In short, one approach—the hard path—uses the new genetic science to engineer radical changes in the very blueprint of species to advance progress, while the other approach—the soft path—uses the same genetic science to create a more sustainable relationship between existing species and their environments.

It might be asked, in the case of agriculture and medicine and any number of other fields, why both approaches to applied science can't live side by side, each complementing and augmenting the other. In reality, the commercial market favors the more reductionist approach for the obvious reason that for now, at least, that's where the money is to be made. While there is certainly a growing market for organic produce and preventive health practices, programs, and products, far more money is invested in biotech agriculture and "illness"-based medicine. That could change, but it would require a paradigm shift in the way we think about science and its applications, with awareness of and support for a science founded in systems thinking and sensitive to the twin notions of diversity and interdependence.

Each of the visions of science I've outlined are based on different sets of values, although I suspect that most molecular biologists continue to entertain the notion that their approach is unbiased, objective, value-free, and the only true science. Their remonstrances notwithstanding, what you see ultimately depends on what you're looking for. The search is always preconditioned by the biases of the researchers.

Of course, science goes hand in hand with technology. Here again, it is amazing how unexamined has been our approach to the new gene-spliced biotechnologies. The fatalistic attitude that if it can be done, it will be done, speaks volumes to our current understanding of and relationship to technology. We've come to view technological "advances" much the way we view the evolution of nature, as if each is fated and irrepressible, the implicit message being that to oppose their introduction is as ill-advised and futile as opposing nature's own steady advance. I know molecular biologists who sincerely believe that their ability to make changes in the genetic code of living things represents the inevitable next step in the evolutionary process and is as unstoppable as natural selection itself. For more than a century we have also labored under the preposterous but nonetheless deeply held belief that technologies are neutral and value-free. The very notion that technological innovations might be socially constructed projections of a particular world view, nurtured by market forces and made current by the prevailing social milieu, would be unthinkable to most scientists.

By vesting every new technology with neutrality and inevitability, the many special interests who have so much to gain from the speedy introduction and acceptance of their "inventions" free themselves of any responsibility for having to ponder the merit, wisdom, or appropriateness of their "contributions." Technologies, however, are not value-free, nor are they inevitable. The fact is, technologies are amplifications and extensions of our biological bodies, appendages we create out of the stuff of the Earth to help inflate ourselves so that we might more easily overcome spatial limitations, minimize temporal constraints, and better expropriate and consume the world around us. A bow and arrow is an extension of our throwing arm. Automobiles extend our legs and feet. Computers amplify our memory. Every tool we've ever created represents increments of power, a way to exercise an advantage over the forces of nature and each other. The exercise of that power is never neutral, for in the act of utilizing the power inherent in each new tool we fashion, someone or something in the environment is compromised, diminished, or exploited to enhance or secure our own well-being. The point is, power is never neutral. There are always winners and losers whenever power is applied.

The question, then, that should be asked of any new technology being readied for society is whether the power being exercised is appropriate or inordinate in scale or scope. Are there technologies whose inherent power is so immense and overshadowing that the unleashing of that power will result in greater diminution than enhancement and more harm than good? Does the technology ultimately drain rather than sustain the labyrinth of relationships that sustain our lives? Nuclear power is a good example of a technology whose inherent power is so utterly overwhelming and beyond appropriate scale that it inflicts more harm than good. The ever-present danger of nuclear meltdowns and the long-term threat of accumulating radioactive wastes makes any short-term market benefits pale in comparison. Nuclear power drains rather than sustains the environment and, therefore, ought not be used as an energy source.

The splitting of the atom and the unraveling of the DNA double helix represent the two premier scientific accomplishments of the twentieth century, the first a tour de force of physics, the second of biology. Both, when applied in the form of new technologies, represent unparalleled potential power to alter both the physical and natural worlds. In the case of nuclear technology, in the form of the bomb and nuclear energy, some nations belatedly chose to reduce and even discontinue their production and use, concluding that the risk in deployment, both to the environment and current and future generations, exceeded any potential benefits. Only two atomic bombs have been dropped on human populations in more than a half century. Nuclear energy, once considered the greatest source of power ever developed, has been partly or largely abandoned in many countries for financial and environmental reasons. In both cases, it was the public that forced the change in policy. Most of the physicists involved in the research as well as the industries that financed and profited from the development and introduction of these two powerful nuclear technologies continue to champion their development to this very day.

If the century just passing was the age of physics and nuclear technology its crown jewel, then the century just coming into view will belong to biology and its premier technology will be genetic engineering. It seems altogether reasonable then, on the cusp of this new century, to ask the critical threshold question that ought to be asked of any new technology revolution. Is the power inherent to the new genetic technologies an appropriate exercise of power? Does it preserve and enhance rather than destabilize and deplete the biological diversity of the planet? Is it easily manageable or ultimately uncontrollable? Does it protect the options or narrow the opportunities for future generations and the other creatures who travel with us? Does it promote respect for life or diminish it? On balance, does it do more good than harm?

While it might seem highly improbable, even inconceivable, to most of the principal players in this new technology revolution that genetic engineering, with all of its potential promise, might ultimately be partially rejected, we need remind ourselves that just a generation ago, it would have been just as inconceivable to imagine the partial abandonment of nuclear energy which had for years been so enthusiastically embraced as the ultimate salvation for a society whose appetite for energy appeared nearly insatiable. It is possible that society will accept some and reject other uses of genetic engineering in the coming Biotech Century. For example, one could make a solid case for genetic screening—with the appropriate safeguards in place—to better predict the onslaught of disabling diseases, especially those that can be prevented with early treatment. The new gene-splicing technologies also open the door to a new generation of life-saving pharmaceutical products. On the other hand, the use of gene therapy to make corrective changes in the human germ line, affecting the options of future generations, is far more problematic, as is the effort to release large numbers of transgenic organisms into the Earth's biosphere. Society may well say yes to some of the genetic engineering options and no to others. After all, nuclear technology has been harnessed effectively for uses other than creating energy and making bombs.

Even a rejection of some genetic engineering technologies does not mean that the wealth of genomic and environmental information being collected couldn't be used in other ways. While the twenty-first century will be the Age of Biology, the technological application of the knowledge we gain can take a variety of forms. To believe that genetic engineering is the only way to apply our newfound knowledge of biology and the life sciences is limiting and keeps us from entertaining other options which might prove even more effective in addressing the needs and fulfilling the dreams of current and future generations.

It needs to be stressed that it's not a matter of saying yes or no to the use of technology itself and never has been—although many in the scientific establishment like to frame the issue this way, leaving the impression that if one is opposed to their particular technological vision, one is anti-technology. In this sense their position on technology mirrors their position on science, in both cases taking a fundamentalist view that there is only one "true path" to the future.

Rather, the question is what kind of biotechnologies will we choose in the coming Biotech Century? Will we use our new insights into the workings of plant and animal genomes to create genetically engineered "super crops" and transgenic animals, or new techniques for advancing ecological agriculture and more humane animal husbandry practices? Will we use the information we're collecting on the human genome to alter our genetic makeup or to pursue new sophisticated health prevention practices?

Human beings are tool makers by nature. We are continually rearranging and altering our environment to secure our well-being and enhance our prospects for a better way of life. We are also risk takers. How then, do we know which tools to use and what risks to take? Since it is impossible to be clairvoyant and know all of the potential ramifications and consequences that might accompany the many new technologies we might want to introduce, we should attempt to minimize regrets and keep open as many options as possible for those who come after us—including our fellow creatures. This means that when choosing among alternative technological applications, we are best served by taking the less radical, more conservative approach—the one least likely to create disruptions and externalities. "First, do no harm" is a well-established and long revered principle in medicine. The fact is, the more powerful a technology is at altering and transforming the natural world—that is, marshaling the environment for immediate, efficient, and short-term ends—the more likely it is to disrupt and undermine long-standing networks of relationships and create disequilibrium somewhere else in the surrounding milieu. Which of the two competing visions of biotechnology—genetic engineering or ecological practices and preventive health—is more radical and adventurous and most likely to cause disequilibrium and which is the more conservative approach and least likely to cause unanticipated harm down the line? The answer, I believe, is obvious.

We may decide, in the final analysis, to shift technological priorities altogether. Now, the genetic engineering technologies are the dominant mode of application of the new biological sciences. The more integrative and embedded technological applications, the ones more sensitive to ecosystem dynamics and interrelationships, remain marginal to the unfolding of the Biotech Century. However, it is not difficult to imagine a turnaround of sorts, in the years ahead, with the more ecological and sustainable biotechnologies taking precedence, and with some genetic engineering technologies being abandoned and others used in a limited fashion and only as options of last resort. For example, in those cases where prevention and holistic health practices are insufficient to ward off seriously debilitating or deadly genetic diseases, somatic gene surgery may be an appropriate remedy.

We should also consider the very real possibility that the new genetic engineering technologies may not, in the final analysis, deliver on many of their promises. The reason I say this is because most molecular biologists, while they use the language of the new cosmological narrative, are still wed to the older, industrial frame of mind. They continue to try to force living processes into linear contexts, believing it possible to manipulate development, gene by gene, as if an organism were merely an assemblage of the individual genes that make it up. This old-fashioned reductionist approach to biotechnology, with its emphasis on sequentiality and strict causality, is likely to meet with only limited success. The Biotech Century will ultimately belong to the systems thinkers, those who see biology more as "process" than "construction" and who view the gene, the organism, the ecosystem, and the biosphere as an integrated "super organism," with the health of each part dependent on the health and well-being of the whole system. That is why the genetic engineers might eventually lose their dominant position to the ecologists whose thinking is more in tune with a biosphere consciousness. If that were to happen, alternative biotechnologies might yet triumph over gene-splicing techniques in the Biotech Century.

Unfortunately, most of the discussion of the biotech revolution, thus far, has centered less on the weighty issues at hand, and more on the "claims" of the scientists conducting the research and the "motivations" of the critics challenging their work. The stereotyping on both sides of the debate has done little to advance the larger questions that need to be asked as we journey into the Biotech Century. My own sense is that most of the molecular biologists engaged in gene research are motivated as much by their desire to make a meaningful contribution to science and enhance the human condition as they are by dreams of financial rewards. Similarly, while most of the companies pursuing this new economic frontier are driven by the prospect of reaping commercial gain, I have talked to enough business leaders in the biotech industry to know that they too believe that their efforts will improve the lot of life for millions of people and make the world a better and more secure place for future generations.

The fact is, many of the new products and processes of the fledgling biotech revolution are of potential benefit. If they weren't, they wouldn't find a commercial market. Companies aren't in business to make products and provide services people don't want. And that's exactly the point. The issue is not simply the motivation of the scientists or the companies financing the research, but, rather, the motivation of the rest of us whose expectations, desires, attitudes, and biases set the cultural parameters for the kind of future we chart as a civilization.

Some will argue that it's not as simple as saying we get the future we want and expect. After all, most people have little control over the kind of research being pursued and even less ability to influence decisions made in corporate boardrooms over which kinds of products and services should be produced and marketed. Nor do most people have any effective way of countering or turning away from the barrage of mass media and advertising that are such pervasive forces in shaping societal values. All of this is true. Still, consumers create markets as much as markets create consumers. Despite the overwhelming pressure of these institutional forces, in the final analysis, I believe that each of us is responsible, in some way, for determining the collective future we share together as a species. To think otherwise would be to suggest that most of us are little more than passive observers of our own destinies, our fate always in someone else's hands. In some ways, it's easier to think that way, since it absolves each of us from having to take responsibility for the world we inherit, experience, and pass on.

The biotechnology revolution will affect each of us more directly, forcefully, and intimately than any other technology revolution in history. For that reason alone every human being has a direct and immediate stake in the direction biotechnology will take in the coming century. Until now, the debate over biotechnology has engaged a narrow group of molecular biologists, industry executives, government policy makers, and critics. With the new technologies flooding into the marketplace and into our lives, the moment has arrived for a much broader debate over the benefits and risks of the new science, one that extends beyond professional authorities and "experts" on both sides of the issue and includes the whole of society. The discussion will need to be as deep as it is broad. The biotech revolution raises fundamental questions about the nature of science, the kinds of new technologies we introduce into the marketplace, and the role of commerce in the intimate affairs of biology.

I have shared some of my own thoughts, opinions, and biases, as well as those of other critics, in the preceding pages. They represent only one point of view. Many molecular biologists and industry leaders in this new field hold quite different views, fervently adhered to and buttressed by their own compelling arguments. My hope is that we might now invite the rest of society to come to the fore for a rich and robust conversation over the kind of future we'd like for ourselves, our children, and the other creatures with whom we share this planet.

Skeptics will say it's naive to believe that most people either care about or desire to participate in "abstract" issues far removed from their day-to-day lives. Yet, the questions surrounding the new technologies are neither abstract nor remote. Quite the contrary, they are the most intimate and pressing ever to face humanity and are of concern to every human being living on Earth. This point came home to me recently on a visit to a medium-size city, Ribeirão Prêto, deep in the interior of Brazil. My hosts, none of whom were involved in the biotech revolution, told me that the news of the birth of Dolly, the cloned sheep, had come as a shock and became a topic of intense discussion and debate among local farmers and townspeople. Friends and neighbors wondered out loud in beer halls, at social gatherings, and around kitchen tables about the implications of this startling new development and sought to understand its potential benefits and risks and how it might affect both their lives and their children.

The biotech revolution will affect every aspect of our lives. The way we eat; the way we date and marry; the way we have our babies; the way our children are raised and educated; the way we work; the way we engage in politics; the way we express our faith; the way we perceive the world around us and our place in it—all of our individual and shared realities will be deeply touched by the new technologies of the Biotech Century. Surely, these very personal technologies deserve to be widely discussed and debated by the public at large before they become a part of our daily lives.

The biotech revolution will force each of us to put a mirror to our most deeply held values, making us ponder the ultimate question of the purpose and meaning of existence. This may turn out to be its most important contribution. The rest is up to us.

# Notes

## Introduction

1. Weinberg, Janet H., "Decision at Asilomar," *Science News,* March 22, 1975, p. 196.
2. Rogers, Michael, "The Pandora's Box Congress," *Rolling Stone,* June 19, 1975, p. 77.

## One **The Biotech Century**

1. *World Almanac and Book of Facts, 1997* (Mahwah, NJ: World Almanac Books, 1996), p. 553; *Statistical Yearbook, 1957* (New York: Statistical Office of the United Nations, Department of Economic and Social Affairs, 1957), p. 35.
2. Mumford, Lewis, *The Myth of the Machine: Technics and Human Development* (New York: Harcourt, Brace & World, 1967), p. 124.
3. Wertime, Theodore, and James D. Muhly, eds., *The Coming of the Age of Iron* (New Haven: Yale University Press, 1980), p. 9.
4. Quoted in Erik Eckholm, "Disappearing Species: The Social Challenge," Worldwatch Paper 22 (Washington, D.C.: Worldwatch Institute, June, 1978), p. 6; Wilson, Edward O., *The Diversity of Life* (Cambridge, MA: Harvard University Press, 1992), p. 280.
5. Bishop, Jerry E., and Michael Waldholz, *Genome: The Story of the Most Astonishing Scientific Adventure of Our Time—The Attempt to Map All the Genes in the Human Body* (New York: Simon & Schuster, 1990), p. 203.
6. Ibid., p. 210.
7. Ibid., pp. 213–14.
8. Stipp, David, "Genetic Map That Could Speed Diagnosis of Inherited Disease Touches Off Dispute," *Wall Street Journal,* November 8, 1987; Collaborative Research Inc., Bedford, MA, *Annual Report for 1987,* p. 3.
9. McKusick, Victor A., "Mapping and Sequencing the Human Genome," *New England Journal of Medicine,* vol. 320, 1989, pp. 912–13.
10. Schuler, D., et al., "A Gene Map of the Human Genome," *Science,* vol. 274, October 25, 1996, p. 540.
11. Lord Richie-Calder, "Retailoring the Tailor," *1976 Encyclopaedia Britannica, Book of the Year* (London: William Benton, 1975), Special Supplement, p. iv.
12. United States Office of Technology Assessment, *Impacts of Applied Genetics* (Washington, D.C.: U.S. Government Printing Office, 1981), p. 8.
13. Hotz, Robert Lee, and Thomas H. Maugh II, "Biotech: The Revolution Is Already Underway," *Los Angeles Times,* April 27, 1997, p. A28.
14. Carey, John, Naomi Freudlich, Julia Flynn, and Neil Gross, "The Biotech Century," *BusinessWeek,* March 10, 1997, p. 79.
15. O'Toole, Thomas, "In the Lab: Bugs to Grow Wheat, Eat Metal," *Washington Post,* June 18, 1980, p. A1.
16. Frederick, Robert J., and Margaret Egan, "Environmentally Compatible Applications of Biotechnology: Using Living Organisms to Minimize Harmful Human Impact on the Environment," *BioScience,* Spring 1994, p. 531.

17. Ibid.

18. Ibid., p. 532.

19. Ibid.; Carey et al., "The Biotech Century," p. 88.

20. Ball, Philip, "Living Factories," *New Scientist,* February 3, 1996, pp. 28–31.

21. Frederick and Egan, "Environmentally Compatible Applications of Biotechnology," p. 530.

22. Carey et al., "The Biotech Century," p. 88.

23. Parkin, Gene F., "Bioremediation: A Promising Technology," in Frederick B. Rudolph and Larry V. McIntire, eds., *Biotechnology: Science, Engineering, and Ethical Challenges for the Twenty-First Century* (Washington, D.C.: Joseph Henry Press, 1996), p. 117.

24. Carey et al., "The Biotech Century," p. 88.

25. Hotz and Maugh, "Biotech: The Revolution Is Already Underway," p. A28.

26. Weiss, Rick, "Mutant Bugs: Genetically Altered Heroes or Spineless Menaces?" *Washington Post,* December 18, 1995, p. A3.

27. Busch, Lawrence, William B. Lacy, Jeffrey Burckhardt, and Laura R. Lacy, *Plants, Power, and Profit: Social, Economic, and Ethical Consequences of the New Biotechnologies* (Cambridge, MA: Basil Blackwell, 1991), p. 173. See Rogoff, Martin H., and Stephen L. Rawlins, "Food Security: A Technological Alternative," *BioScience,* December 1987, pp. 800–807.

28. "Tricking Cotton to Think Lab Is Home Sweet Home," *Washington Post,* May 29, 1988, p. A3.

29. Rogoff and Rawlins, "Food Security," pp. 800–807; interview, May 11, 1994. Stephen Rawlins says that in the coming era of highly automated laboratory farming, the only part of the process that needs to remain outdoors is capturing the energy of the sun in biomass plants. "You have to capture the energy outdoors because that's where the sun is. But the rest of the processes, once you have the energy, don't have to be outdoors." Rawlins adds that "by bringing [farming] indoors . . . you don't have all of the environmental problems."

30. Ford, Jane, "This Little Pig Rushed to Market," *New Scientist,* April 28, 1988, p. 27.

31. Cooney, Bob, "Antisense Gene Could Knock out Broodiness in Turkeys," *Science Report,* Agricultural and Consumer Press Service, College of Agricultural and Life Sciences, Research Division, University of Wisconsin—Madison, May 4, 1993; Hoffman, Michelle, "Building a Badder Mother," *American Scientist,* July/August 1993, p. 329. See B. Wentworth, H. Tsai, A. Wentworth, E. Wong, J. Proudman, and M. El Halawani, "Primordial Germ Cells for Genetic Modification of Poultry," in R. H. Miller, V. G. Pursel, and H. D. Norman, eds., *Biotechnology's Role in the Genetic Improvement of Farm Animals* (Savoy, IL: American Society of Animal Science, 1996), pp. 202–27.

32. Thayer, Ann, "Firms Boost Prospects for Transgenic Drugs," *Chemical and Engineering News,* August 26, 1996, pp. 23–24; Johannes, Laura, "Biotech Goat Is Created to Produce Drug," *Wall Street Journal,* April 9, 1997, pp. B1, B6.

33. Graves, Martha, "Transgenic Livestock May Become Biotech's Cash Cow," *Los Angeles Times,* May 1, 1997, p. A12.

34. Thayer, "Firms Boost Prospects for Transgenic Drugs," p. 23.

35. Carey et al., "The Biotech Century," pp. 79–80.

36. Kolata, Gina, "Lab Yields Lamb with Human Gene," *New York Times,* July 25, 1997, p. A18.

37. Graves, "Transgenic Livestock May Become Biotech's Cash Cow," p. A12.

38. Ibid.

39. Smith, Emily T., Andrea Durham, Edith Terry, Neil Gilbride, Phil Adamsak, and Jo Ellen Davis, "How Genetics May Multiply the Bounty of the Sea," *BusinessWeek,* December 16, 1985, pp. 94–95.

40. Ibid., p. 95.

41. Holmes, Bob, "Blue Revolutionaries," *New Scientist,* December 1996, p. 32.

42. Hotz and Maugh, "Biotech: The Revolution Is Already Underway," p. A28.

43. Weiss, "Mutant Bugs: Genetically Altered Heroes of Spineless Menaces?" p. A3.

44. Grady, Denise, "A Lab Breeds a Mighty Mouse, with a Variety of Implications," *New York Times*, May 1, 1997, p. B14; Weiss, Rick, "Mighty Mice Are On the Way: Genetic Altering Increases Muscles, Not Fat," *Washington Post*, May 1, 1997, pp. A1, A15.

45. Weiss, Rick, "Human Chromosome Transplanted Into Mice: Potential Is Seen for Disease-Fighting Antibodies," *Washington Post*, May 30, 1997, pp. A1, A6.

46. Edelson, Edward, "Body Builders," *Popular Science*, May 1996, p. 61.

47. Langer, Robert, and Joseph P. Vacanti, "Artificial Organs," *Scientific American*, September 1995, p. 130.

48. Ibid., pp. 131–32.

49. Ibid., pp. 62–63.

50. Langer and Vacanti, "Artificial Organs," p. 132.

51. Ibid., pp. 131–32; Edelson, "Body Builders," pp. 62–64; Dobson, Roger, "Fabricating Human Organs Is No Longer Science Fiction," *The Sunday Times of London*, November 23, 1997, p. 1.

52. Langer and Vacanti, "Artificial Organs," p. 132.

53. Ibid.

54. Wade, Nicholas, "Rapid Gains Are Reported on Genome: Sequencing May Be 99% Done by 2002," *New York Times*, September 28, 1995, p. A24.

55. Eng, Charis, and Jan Vijg, "Genetic Testing: The Problems and the Promise," *Nature Biotechnology*, May 1997, p. 426; Carey et al., "The Biotech Century," p. 85.

56. Weiss, Rick, "Artificial Human Chromosomes That Replicate Developed in Lab: Scientists Aim to Ferry Curative Genes to Cells," *Washington Post*, April 1, 1997, pp. A1, A6.

57. Ibid.

58. Halacy Jr., D.S., *Genetic Revolution: Shaping Life for Tomorrow* (New York: Harper & Row, 1974), p. 148; Rosenfeld, Albert, *The Second Genesis: The Coming Control of Life* (New York: Vintage, 1969), p. 175.; U.S. Congress Office of Technology Assessment, *Artificial Insemination Practice in the United States: Summary of a 1987 Survey—Background Paper*, OTA-BP-BA-48 (Washington, DC: U.S. Government Printing Office, August 1988), p. 40; Kolata, Gina, "Clinics Selling Embryos Made for Adoption," *New York Times*, November 23, 1997, p. A1.

59. Kotulak, Ronald, and Peter Gorner, "Babies By Design," *Chicago Tribune*, March 3, 1991, p. C14.

60. Wallis, Claudia, "The New Origins of Life: How the Science of Conception Brings Hope to Childless Couples," *Time*, September 10, 1984, p. 46.

61. *The Business of Surrogate Parenting* (Albany, NY: New York State Department of Health, April 1992), p. 3.

62. Fletcher, Joseph, *The Ethics of Genetic Control: Ending Reproduction Roulette* (Garden City, NY: Anchor Books, 1974), p. 103.

63. Cited in Albert Rosenfeld, *The Second Genesis: The Coming Control of Life* (New York: Vintage, 1969), p. 139.

64. Langer and Vacanti, "Artificial Organs," p. 133.

65. Connor, Steve, and Deborah Cadbury, "Headless Frog Opens Way for Human Organ Factory," *The Sunday Times* (London), October 19, 1997.

66. Ibid.

67. Ibid.

68. Quoted in *Bioworld Today*, June 12, 1996.

69. Berman, Morris, *The Reenchantment of the World* (Ithaca, NY: Cornell University Press, 1981), p. 92.

70. Ibid., p. 88.

71. Quoted in Titus Burckhardt, *Alchemy: Science of the Cosmos, Science of the Soul*, William Stoddart, translator (London: Stuart & Watkins, 1967), p. 25.

72. Eisner, Thomas, "Chemical Ecology and Genetic Engineering: The Prospects for Plant Protection and the Need for Plant Habitat Conservation," *Symposium on Tropical Biology and Agriculture* (St. Louis, MO: Monsanto Company, July 15, 1985).

## Two **Patenting Life**

1. World Commission on Environment and Development, *Our Common Future [The Brundtland Report]* (Oxford: Oxford University Press, 1987), p. 155; Fisher, A. C., "Economic Analysis and the Extinction of Species," Department of Energy and Resources, University of California, Berkeley, 1982.

2. Gary Nabhan has conducted an extensive survey on the loss of seed varieties (germplasm) in North America. His work *Enduring Seeds: Native American Agriculture and Wild Plant Conservation* (San Francisco: North Point Press, 1989) documents the evolution of seed propagation in the United States and warns of the potential dangers in losing this important natural resource. According to most estimates, roughly half of all our farm animal species face extinction in the coming years. See, for example, Lisa Drew, "The Barnyard Restoration," *Newsweek*, May 29, 1989, pp. 50–51. For literature on the availability of extremely rare or "minor" animal breeds, contact the American Minor Breeds Conservancy, Box 477, Pittsboro, NC, 27312.

3. Slater, Gilbert, *The English Peasantry and the Enclosure of Common Fields* (New York: A. M. Kelley, 1968), p. 1. The literature on the enclosure movement is voluminous; however, the history of enclosure is generally not well known in American academic circles. The best sources on the enclosure movement in Britain are: Tawney, R. H., *The Agrarian Problem in the Sixteenth Century* (London: Longmans Green, 1912); Tate, William, *The English Village Community and the Enclosure Movement* (New York: Walker, 1967); Hammond, John L., and Barbara B. Hammond, *The Village Labourer, 1790–1832: A Study in the Government of England Before the Reform Bill* (London: Longmans Green, 1911); Finberg, H. P. R., and Joan Thirsk, eds., *The Agrarian History of England and Wales, 1500–1640*, vol. 4 (Cambridge: Cambridge University Press, 1967).

4. Polanyi, Karl, *The Great Transformation: The Political and Economic Origins of Our Time* (Boston: Beacon Press, 1957), p. 35; Rubenstein, Richard, *The Age of Triage: Fear and Hope in an Overcrowded World* (Boston: Beacon Press, 1983), p. 10.

5. Rubenstein, *The Age of Triage*, p. 43; Slater, *The English Peasantry and the Enclosure of Common Fields*, pp. 6, 110.

6. Slater, *The English Peasantry and the Enclosure of Common Fields*, p. 4.

7. *Diamond, Sidney A., Commissioner of Patents and Trademarks, petitioner, v Chakrabarty, Ananda M., et al.*, 65 L ed 2d 144, June 16, 1980, p. 148.

8. See *In re Bergy*, 563 F. 2d 1031 (1975).

9. Howard, Ted, "The Case Against Patenting Life," brief on behalf of the People's Business Commission, Amicus Curiae, in the Supreme Court of the United States, no. 79–136, p. 29.

10. *Diamond v Chakrabarty*, p. 152.

11. Ibid., p. 158.

12. Ibid., p. 154.

13. Quoted in Sharon McAuliffe and Kathleen McAuliffe *Life for Sale* (New York: Coward, McCann and Geoghegan, 1981), p. 11.

14. Ibid., p. 28.

15. Ibid., p. 205.

16. Kass, Leon R., "Patenting Life," *Commentary*, December 1981, p. 56.

17. U.S. Patent and Trademark Office, *Animals—Patentability* (Washington, D.C.: U.S. Government Printing Office, April 7, 1987).

18. Powledge, Fred, "Who Owns Rice and Beans?" *BioScience*, July/August 1995, p. 442.

19. Quoted in Brian Belcher and Geoffrey Hawtin, *A Patent on Life: Ownership of Plant and Animal Research* (Canada: IDRC, 1991).

20. Raines, Lisa, "Of Mice and Men and Tennis Balls," *Across the Board*, March 1989, p. 46.

21. Testimony of Michael Glough, U.S. Congress, Office of Technology Assessment, Before

the Subcommittee on Intellectual Property and Judicial Administration House Committee on the Judiciary, November 20, 1991, on "Patents and Biotechnology," p. 3.

22. U.S. Congress, Office of Technology Assessment, *New Developments in Biotechnology: Patenting Life—Special Report, OTA-BA-370* (Washington, D.C.: U.S. Government Printing Office, April 1989), p. 37.

23. Powledge, "Who Owns Rice and Beans?" p. 440.

24. Ibid., p. 441.

25. Quoted in Hope Shand, "Patenting the Planet," *Multinational Monitor,* June 1994, p. 13.

26. Crosby, Alfred W., Jr., *The Columbian Exchange: Biological and Cultural Consequences of 1492* (Westport: Greenwood Publishing Co., 1972); Kloppenburg, Jack R., Jr., *First the Seed: The Political Economy of Plant Biotechnology, 1492–2000* (Cambridge: Cambridge University Press, 1988); Mooney, Pat R., "The Law of the Seed: Another Development and Plant Genetic Resources," *Development Dialogue,* vols. 1 and 2, 1983; Brockway, Lucile H., *Science and Colonial Expansion: The Role of the British Royal Botanic Gardens* (New York: Academic Press, 1979); *The Encyclopaedia Britannica,* vol. 19 (London: William Benton, 1969), p. 682; Reed, Howard S., *A Short History of the Plant Sciences* (Waltham, MA: Chronica Botanica Co., 1942).

27. Shand, "Patenting the Planet," p. 10; Sears, Cathy, and Robert Nexworth, "Drug Wars: Extracting Patents and Profits from the Rainforests' Medicinal Plants," *Village Voice,* July 21, 1992, p. 39.

28. Shiva, Vandana, *Biopiracy: The Plunder of Nature and Knowledge* (Boston: South End Press, 1997), p. 69; Hirsh, Michael, "Fight for the Miracle Tree," *Bulletin,* September 26, 1995, pp. 70–71.

29. National Research Council, "Neem, A Tree for Solving Global Problems," *National Research Council Report* (Washington, D.C.: National Academy Press, 1995), p. 5.; Stone, Richard, "A Biopesticidal Tree Begins to Blossom," *Science,* February 28, 1992, p. 1070.

30. Letter from B. N. Dhawan of the Central Drug Research Institute to Dr. Vandana Shiva, Director of the Research Foundation for Science, Technology, and National Resource Policy, June 14, 1993, cited in Foundation on Economic Trends Petition to U.S. Patent and Trademark Office for Request for Reexamination on U.S. Patent No. 5,124,349, issued June 23, 1992, for Storage Stable Azadirachtin Formulations.

31. Enyart, James, "A GATT Intellectual Property Code," *Les Nouvelles,* June 1990, pp. 54–56.

32. Shiva, *Biopiracy,* p. 10.

33. Powledge, "Who Owns Rice and Beans?" p. 443.

34. Shand, "Patenting the Planet," p. 12.

35. Putterman, Daniel, "Compromise Sought over Germplasm Access," *Nature,* November 11, 1994, p. 9.

36. Powledge, "Who Owns Rice and Beans?" p. 444; Tangley, Laura, "Ground Rules Emerge for Marine Prospectors," *BioScience,* April 1996, p. 246.

37. Tangley, "Ground Rules Emerge for Marine Prospectors," p. 246.

38. Ibid; Shiva, *Biopiracy,* p. 75.

39. Kenney, Martin, *Biotechnology: The University-Industrial Complex* (New Haven: Yale University Press, 1986), p. 110.

40. Blumenthal, David, et al., "Industrial Support of University Research in Biotechnology," *Science,* January 17, 1986, pp. 242–46.

41. Blumenthal, David, et al., "University-Industry Research Relationships in Biotechnology: Implications for the University," *Science,* June 13, 1986, pp. 1361–66.

42. Ibid.

43. Krimsky, Sheldon, *Biotechnics and Society: The Rise of Industrial Genetics* (New York: Praeger, 1991), p. 77.

44. Yamamoto, Keith R., "Faculty Members as Corporate Officers: Does Cost Outweigh Benefits?" in William J. Whelan and Sandra Black, eds., *From Genetic Experimentation to Biotechnology: The Critical Transition* (Chichester, England: Wiley, 1982), p. 198.

45. Lehrman, Sally, "Diversity Project: Cavalli-Sforza Answers His Critics," *Nature,* May 2, 1996, p. 14.

46. Bright, Chris, "Who Owns Indigenous Peoples' DNA?" *Humanist,* January 1995, p. 44.

47. Ibid.; Shand, Hope, "Extracting Human Resources," *Multinational Monitor,* June 1994, p. 11.

48. Shand, "Extracting Human Resources," p. 11.

49. Lehrman, Sally, "Anthropologist Cleared in Patent Dispute," *Nature,* April 4, 1996, p. 374.

50. Dickson, David, "Whose Genes Are They Anyway?" *Nature,* May 2, 1996, p. 13.

51. Jayaraman, K. S., "Indian Researchers Press for Stricter Rules to Regulate 'Gene Hunting,' " *Nature,* February 1, 1996, p. 381.

52. Ibid.

53. Jayaraman, K. S., "Gene-Hunters Home In on India," *Nature,* May 2, 1996, p. 13.

54. Jayaraman, "Indian Researchers Press for Stricter Rules to Regulate 'Gene-Hunting,' " p. 382.

55. Jayaraman, K. S., and Colin Macilwain, "Scientists Challenged over Unauthorized Export of Data," *Nature,* February 1, 1996, p. 381.

56. Jayaraman, "Gene-Hunters Home In on India," p. 13.

57. Dickson, "Whose Genes Are They Anyway?" p. 11.

58. Monmaney, Terence, "Gene Sleuths Seek Asthma's Secrets on Remote Island," *Los Angeles Times,* April 30, 1997, p. A12.

59. *Moore v The Regents of the University of California, et al.,* Supreme Court of the State of California, p. 23.

60. Koechlin, Florianne, "No Patents on Life! Mail-Out 48," No Patents on Life!, European Coordination, February/March 1997, p. 2.

61. See United States Patent #5,061,620, October 29, 1991, "Human Hematopoietic Stem Cell"; Waldholz, Michael, and Hilary Stout, "A New Debate Rages over the Patenting of Gene Discoveries," *Wall Street Journal,* April 17, 1992, p. A6.

62. Dickson, "Whose Genes Are They Anyway?" p. 11.

63. Roberts, Leslie, "NIH Gene Patents, Round Two," *Science,* February 21, 1992, pp. 912–13; Herman, Robin, "NIH Genes Researcher Is Leaving for His Own Lab," *The Washington Post,* July 7, 1992, Health, p. 4.

64. Waldholz and Stout, "A New Debate Rages over the Patenting of Gene Discoveries," pp. A1, A6.

65. Lewin, Tamar, "Move to Patent Cancer Gene Is Called Obstacle to Research," *New York Times,* May 21, 1996, p. A14.

66. Emmot, Steve, "The Directive Rises Again," *Seedling,* March 1997.

67. Thoenes, Sander, "EU Lobbyists Shout Louder to Be Heard," *Financial Times,* July 18, 1997, p. 3.

68. de Chobam, Thomas, *Summa Confessorum,* F. Broomfield, ed. (Paris: Louvain, 1968), p. 505.

69. Rosewicz, Barbara, and Michael Waldholz, "Human, Animal Gene Patents Targeted by a Religious Coalition in a Petition," *Wall Street Journal,* May 15, 1995, p. B2.

## Three **A Second Genesis**

1. Rural Advancement Foundation International, "The Life Industry," *RAFI Communique,* September 1996, p. 1, <http://www.rafi.ca/rafi/communique/fltxt/19964.html> (September 23, 1997). In addition to the resources provided by RAFI's website, reference material covering the life science industry, pp. 93–97, was collected from the following sources: Monsanto Company, "Monsanto Reaches Agreements to Acquire Holden's Corn States and Corn States International; Complementary Technologies Important to Monsanto's Life Science Business," *In the News: Press Releases,* 1997, <http://www.monsanto.com/monpub/in the news/press releases/> (September 23, 1997);

Asgrow Company, "What's New at Asgrow," 1996, <http://www.asgrow.com/asgrowcompany/> (September 23, 1997); DeKalb Company, *DeKalb Press Releases,* 1997, <http:www.dekalb.com/> (September 23, 1997), DowElanco Company, "About DowElanco," August 5, 1997, <www.dowelancocom/company.htm/> (September 23, 1997); Mycogen, *News Releases,* 1997, <http:www.mycogen.com/> (September 23, 1997); Du Pont Company, "Du Pont Collaborations in Research and Development," 1995, 1996, <http:www.dupont.com/corp/r-and-d/> (September 23, 1997); Pioneer Hi-Bred Company, "Pioneer and Du Pont Complete Alliance Agreement," September 18, 1997, <http:www.pioneer.com/usa/> (September 23, 1997); Ralston Purina Company, "Ralston Purina Company Intends to Sell Protein Technologies International to Dupont," August 22, 1997, <http:www.ralston.com/news27.html/> (September 23, 1997); Novartis, "Ciba-Geigy Ltd. and Sandoz Ltd. Announce Plans to Merge," March 7, 1996, <http://www.novartis.com/media/index.html> (October 15, 1997).

2. Davidson, Sylvia, "Hidden Biotechnology Worth over $7.5 Billion a Year," *Nature Biotechnology,* May 1996, p. 564.

3. Rural Advancement Foundation International, "The Life Industry," p. 4.

4. Ibid., p. 2.

5. Johnson, Emma, "Gene Therapy Patent Challenge: Round One," *Nature Biotechnology,* April 1996; Coughlan, Andy, "Sweeping Patent Shocks Gene Therapists," *New Scientist,* April 1, 1995, p. 4.

6. Rural Advancement Foundation International, *Life Industry Update: Seeds, Biotech, Agrochemicals,* 1997.

7. Rural Advancement Foundation International, "The Life Industry," pp. 7–8; RAFI maintains an updated website on the life science industry. They can be reached at http://www.rafi.ca/

8. Marx, Jean, "Concerns Raised About Mouse Models for AIDS," *Science,* February 16, 1990, p. 809; Lusso, Paolo, Fulvia Di Marzo Veronese, Barbara Ensoli, Genoveffa Franchini, Cristina Jemma, Susan E. DeRocco, V. V. Kalyanaraman, and Robert C. Gallo, "Expounded HIV 1 Cellular Tropism by Phenotypic Mixing with Murine Endogenous Retroviruses," *Science,* February 16, 1990, p. 851.

9. Rollin, Bernard E., *The Frankenstein Syndrome: Ethical and Social Issues in the Genetic Engineering of Animals* (New York: Cambridge University Press, 1995), pp. 118–19.

10. Ibid., p. 119.

11. Tiedje, J. M., et al., "The Planned Introduction of Genetically Engineered Organisms: Ecological Considerations and Recommendations," *Ecology,* February 1989, p. 302.

12. Hallerman, Eric M., and Anne R. Kapuscinski, series of articles in *Fisheries* ("Transgenic Fish and Public Policy: Anticipating Environmental Impacts of Transgenic Fish," "Transgenic Fish and Public Policy: Regulatory Concerns," and "Transgenic Fish and Public Policy: Patenting of Transgenic Fish"), January 1990, pp. 2–25, as cited in Rollin, *The Frankenstein Syndrome,* p. 123.

13. Mayer, Sue, "Environmental Threats of Transgenic Technology," in Peter Wheale and Ruth McNally, eds., *Animal Genetic Engineering: Of Pigs, Oncomice, and Men* (London: Pluto Press, 1995), p. 128.

14. Lindow, Steven E., "Methods of Preventing Frost Injury Caused by Epiphytic Ice-Nucleation-Active Bacteria," *Plant Disease,* March 1983, pp. 327–33; Advanced Genetic Sciences, "Proposal to Field Test Genetically Engineered Pseudomonas Strains Containing Artificially Introduced Deletions in Ice-Nucleation Genes," submitted to the National Institutes of Health, Recombinant DNA Advisory Committee, March 22, 1984.

15. Odum, Eugene P., "Biotechnology and the Biosphere," *Science,* September 27, 1985, p. 1338.

16. Lindow, Steven E., "Methods of Preventing Frost Injury," p. 332.

17. Advanced Genetic Sciences, "Proposal to Field Test Genetically Engineered Pseudomonas Strains."

18. Snow, Allison A., and Pedro Morán Palma, "Commercialization of Transgenic Plants: Potential Ecological Risks," *BioScience,* February 1997, p. 94.

19. Ibid.

20. Ibid.

21. Steinbrecher, Dr. Ricarda A., "From Green to Gene Revolution: The Environmental Risks of Genetically Engineered Crops," *Ecologist,* November/December 1996, p. 277.

22. Rissler, Jane, and Margaret Mellon, *The Ecological Risks of Engineered Crops* (Cambridge, MA: MIT Press, 1996), pp. 10–11.

23. Ibid., pp. 6, 42–43; Palm, C., K. Donegan, D. Harris, R. Seidler, "Quantification in Soil of *Bacillus thuringiensis* Var. *Kurstaki* Delta-Endotoxin from Transgenic Plants," *Molecular Ecology,* April 1994, pp. 145–51; Union of Concerned Scientists, "Compilation of Data from Applications to the U.S. Department of Agriculture to Field Test Transgenic Crops," Washington, D.C., 1994; U.S. Department of Agriculture, "Scientific Evaluation of the Potential for Pest Resistance to the *Bacillus thuringiensis* (Bt) Delta-Endotoxins," Cooperative State Research Service, Agricultural Research Service, Conference to Explore Resistance Management Strategies, Washington, D.C., 1992; McGaughey, W., and M. Whalon, "Managing Insect Resistance to *Bacillus thuringiensis* Toxins," *Science,* November 27, 1992, pp. 1451–55.

24. Steinbrecher, "From Green to Gene Revolution," p. 273.

25. Goldburg, Rebecca, Jane Rissler, Hope Shand, and Chuck Hassebrook, "Chemical Herbicides and Herbicide-Tolerant Crops," in *Biotechnology's Bitter Harvest* (Biotechnology Working Group, 1990), p. 11.

26. Steinbrecher, "From Green to Gene Revolution," p. 273.

27. Ibid.

28. Ibid.

29. Ibid., pp. 275–76.

30. Rissler and Mellon, *The Ecological Risks of Engineered Crops,* p. 43.

31. Kaiser, Jocelyn, "Pests Overwhelm Bt Cotton Crop," *Science,* July 26, 1996, p. 423; Reifenberg, Anne, and Rhonda L. Rundle, "Buggy Cotton May Cast Doubt on New Seeds," *Wall Street Journal,* July 23, 1996, pp. B1, B3.

32. Quoted in J. L. Fox, "Bt Cotton Infestations Renew Resistance Concerns," *Nature Biotechnology,* September 1996, p. 1070.

33. Rissler and Mellon, *The Ecological Risks of Engineered Corps,* pp. 62–63.

34. Ibid., pp. 35–36.

35. Ibid., p. 41.

36. Ibid., pp. 34–40.

37. Snow and Palma, "Commercialization of Transgenic Plants," p. 91; Rissler and Mellon, *The Ecological Risks of Engineered Crops,* pp. 51–52.

38. Leary, Warren E., "Gene Inserted in Crop Plant Is Shown to Spread to Wild," *New York Times,* March 7, 1996, p. B14.

39. Ibid.

40. Ibid.

41. Steinbrecher, "From Green to Gene Revolution," p. 278; Skogsmyr, I., "Gene Dispersal from Transgenic Potatoes to Cospecifics: A Field Trial," *Theoretic Applications of Genetics,* 1991, pp. 770–71.

42. Weiss, Rick, "Genetically Engineered Rice Raises Fear: As Plants Produce Own Insecticide, Resistance Buildup Could Occur," *The Washington Post,* February 5, 1996, p. A9.

43. U.S. Department of Defense, Biological Defense Program, *Report to the Committee on Appropriations, House of Representatives,* May 1986, p. 4. A brief but informative survey of the development and use of biological and chemical weapons is found in Frank Barnaby, *The Gaia Peace Atlas: Survival Into the Third Millennium* (New York: Doubleday, 1988), pp. 134–38.

44. Ibid., p. 8.

45. Ibid.

46. Ibid., p. 4.

47. Testimony of Douglas J. Feith before the Subcommittee on Oversight and Evaluation of the House Permanent Select Committee on Intelligence, August 8, 1986.

48. Rifkin, Jeremy, *Declaration of a Heretic* (Boston: Routledge & Kegan Paul, 1985), p. 58.

49. Ibid., pp. 58–59.

50. Correspondence from Secretary of Defense Caspar Weinberger to Senator Jim Sasser, November 20, 1984.

51. Tucker, Jonathan B., "Gene Wars," *Foreign Policy,* Winter 1984–85, pp. 60–69; See also Department of Defense Annual Report on Chemical Warfare, Biological Defense Research Program Obligations, October 1, 1984, through September 30, 1985, RCs: DDUSDRE (A) 1065.

52. Horrock, Nicholas, "The New Terror Fear—Biological Weapons: Detecting an Attack Is Just the First Problem," *U.S. News and World Report,* May 12, 1997, p. 36.

53. Dickey, Christopher, "His Secret Weapon—Iraq: Saddam Has a Big Germ-Warfare Arsenal," *Newsweek,* September 4, 1995, p. 34.

54. Cole, Leonard A., "The Specter of Biological Weapons," *Scientific American,* December 1996, p. 62.

55. Langley, Gill, "A Critical View of the Use of Genetically Engineered Animals in the Laboratory," in Wheale and McNally, eds., *Animal Genetic Engineering,* pp. 184–88.

56. Hendricks, Mike, "The Super Pigs Stumble on Road to Pork, Profit," *Kansas City Star,* March 6, 1994, p. A13.

57. Ibid.

58. Ibid.

59. Stevenson, Peter, "Patenting of Transgenic Animals: A Welfare/Rights Perspective," in Wheale and McNally, eds., *Animal Genetic Engineering,* p. 158.

60. McNair, Joel, "BGH Label Lists Potential Health Side Effects," *Agri-View,* November 26, 1993; Official Warning Label for Monsanto's Product *Posilac* (Bovine Growth Hormone) Issued by the United States Food and Drug Administration, April 1993.

61. "FDA Releases Its Own BST Report," *Cheese Market News,* September 23, 1994; "Arguing Till the Cows Come Home," *New Scientist,* October 29, 1994, pp. 14–15; Kastel, Mark, "Down on the Farm: The Real BGH Story—Animal Health Problems, Financial Trouble" (for Rural Vermont, a project of the Rural Education Action Project, Fall 1995); Christiansen, Andrew, "Recombinant Bovine Growth Hormone: Alarming Tests, Unfounded Approval—The Story Behind the Rush to Bring rBGH to Market" (for Rural Vermont, a project of the Rural Education Action Project, July 1995).

62. From proceedings of the NIH Recombinant DNA Advisory Committee (RAC) Spring Meeting, 1985, National Institutes of Health, Bethesda, Maryland, Campus.

63. Ibid.

64. Ibid.

65. Hoban, T. J., and P. A. Kendall, "Consumer Attitudes About the Use of Biotechnology in Agriculture and Food Production," Report to Extension Service, USDA, 1992. See also William K. Hallman and Jennifer Metcalfe, "Public Perceptions of Agricultural Biotechnology: A Survey of New Jersey Residents," Ecosystem Policy Research Center, The New Jersey Agricultural Experiment Station, Cook College, Rutgers, The State University of New Jersey, April 1996; "Public Perceptions of Agri-Biotechnology," *Genetic Engineering News,* July 1995.

66. Ibid.

67. Nordlee, Julie A., Steve L. Taylor, Jeffrey A. Townsend, Laurie A. Thomas, and Robert K. Bush, "Identification of a Brazil-Nut Allergen in Transgenic Soybeans," *New England Journal of Medicine,* March 14, 1996, p. 688.

68. Nestle, Marion, "Allergies to Transgenic Foods: Questions of Policy," *New England Journal of Medicine,* March 14, 1996, p. 726.

69. Ibid.

70. Ibid., p. 727.

71. Samo, Wolfgang, "Pesticides and Agriculture: Industry Perspective," remarks by Novar-

tis' Head of Agribusiness, given at *Government Regulations and Crops Production, An International Conference,* February 24–26, 1997, Tufts University, Boston, Massachusetts, *Agriculture News.*

72. Allan, Jon, "Silk Purse or Sow's Ear," *Nature Medicine,* March 3, 1997, p. 275.

73. Nowak, Rachel, "Xenotransplants Set to Resume," *Science,* November 18, 1994, p. 1148; Allan, Jonathan S., "Fear of Viruses," *New York Times,* January 20, 1996, p. 23.

74. Allan, "Silk Purse or Sow's Ear," pp. 275–76.

75. "Thanks, But No Thanks," *Economist,* October 21, 1995, p. 17.

76. Allan, "Silk Purse or Sow's Ear," p. 276.

77. Kaiser, Jocelyn, "IOM Backs Cautious Experimentation," *Science,* July 19, 1996, p. 305.

78. Allen, William, "Environmental Destruction and Loss of Diversity Require Biotech Attention," *Genetic Engineering News,* July/August 1988, p. 29.

79. Fowler, Cary, and Pat Mooney, *Shattering: Food, Politics, and the Loss of Genetic Diversity* (Tucson, AZ: The University of Arizona Press, 1990), pp. 43–45; Rhoades, Robert, "Incredible Potato," *National Geographic,* May 1982, pp. 679–82.

80. Stakman, E. C., and J. J. Christensen, "The Problem of Breeding Resistant Varieties," in J. G. Horsfall and A. E. Dimond, eds., *Plant Pathology,* vol. 3 (New York: Academic Press, 1960), pp. 574, 578; Browning, J. A., "Corn, Wheat, Rice, Man: Endangered Species," *Journal of Environmental Quality,* July–September 1972, p. 209.

81. Raeburn, Paul, *The Last Harvest: The Genetic Gamble That Threatens to Destroy American Agriculture* (Lincoln, NE: University of Nebraska Press, 1995), p. 167.

82. Ibid., p. 235.

83. Rhoades, "Incredible Potato," p. 694; Myers, Norman, *The Sinking Ark* (Oxford: Pergamon Press, 1979), p. 62; Rick, Charles, quoted in Fowler and Mooney, *Shattering,* p. 77; Rick, Charles, "Conservation of Tomato Species Germplasm," *California Agriculture,* 1977, pp. 32–33; Soria, Jorge, "Recent Cocoa Collecting Expeditions," in O. H. Frankel and J. G. Hawkes, eds., *Crop Genetic Resources for Today and Tomorrow* (New York: Cambridge University Press, 1975), pp. 175–78; Chalmers, W. S., cited in Fowler and Mooney, *Shattering,* p. 78.

84. Raeburn, *The Last Harvest,* p. 167. For further information, contact the Center for Plant Conservation, Missouri Botanical Garden, P.O. Box 299, St. Louis, MO, 63166, (314) 577-9450, over the internet at cpc@mobot.org, or on the World Wide Web at http:\\www.mobot.org\cpc and inquire about their Database of Rare Plant Species.

85. Raeburn, *The Last Harvest,* pp. 168–69.

86. Prescott-Allen, Christine, and Robert Prescott-Allen, *The First Resource: Wild Species in the North American Economy* (New Haven: Yale University Press, 1986), pp. 1, 413.

87. Rural Advancement Foundation International, "Listing of Possible Extinct Varieties of Apples," unpublished report prepared by RAFI, Pittsboro, NC, 1982; Rural Advancement Foundation International, "Listing of Possible Extinct Varieties of Pears," unpublished report prepared by RAFI, Pittsboro, NC, 1982; Fowler and Mooney, *Shattering,* p. 63.

88. Teitel, Martin, *Rain Forest in Your Kitchen: The Hidden Connection Between Extinction and Your Supermarket* (Washington, D.C.: Island Press, 1992), p. 14.

89. Rhoades, Robert E., "The World's Food Supply at Risk," *National Geographic,* April 1991, pp. 74–105.

90. Quoted in Norman Myers, *A Wealth of Wild Species: Storehouse for Human Welfare* (Boulder, CO: Westview Press, 1983), p. 24.

91. "Japan Roundup," *Bio/Technology,* November 1988, p. 1276.

92. Fowler and Mooney, *Shattering,* p. 82; see Erna Bennett, "Historical Perspectives in Genecology," *Scottish Plant Breeding Station Record,* 1964, p. 95.

93. Fowler and Mooney, *Shattering,* pp. 123–24.

## Four **Eugenic Civilization**

1. Roosevelt, Theodore, to Charles B. Davenport, January 3, 1913, "Charles B. Davenport Papers," Department of Genetics, Cold Spring Harbor, NY; Roosevelt, Theodore, "Birth Reform,

from the Positive, Not the Negative Side," in *The Works of Theodore Roosevelt,* vol. 21 (New York: Charles Scribner's Sons, 1923–26), p. 163; Roosevelt, Theodore, "Twisted Eugenics," in *The Works of Theodore Roosevelt,* vol. 12, p. 201.

2. Ludmerer, Kenneth M., *Genetics and American Society: A Historical Appraisal* (Baltimore: Johns Hopkins University Press, 1972), p. 43.

3. Ibid.

4. Guyer, Michael F., *Being Well Born: An Introduction to Heredity and Eugenics* (Indianapolis: Bobbs-Merrill Co., 1916), preface.

5. Conklin, Edwin Grant, "The Future of America: A Biological Forecast," *Harper's Magazine,* April 1928, pp. 529–39.

6. Newman, Horatio H., *Evolution, Genetics and Eugenics* (Chicago: University of Chicago Press, 1921), p. 441.

7. American Breeders Association, *Proceedings,* vol. 2, 1906.

8. Davenport, Charles B., "Charles B. Davenport Papers," January 3, 1913.

9. Haller, Mark H., *Eugenics: Hereditarian Attitudes in American Thought* (New Brunswick, NJ: Rutgers University Press, 1963), p. 73.

10. Davenport, Charles B., "Charles B. Davenport Papers," July 22, 1913.

11. Von Kleinsmid, R. B., "An Inquiry Concerning Some Preventions of Crime," *American Psychological Association,* 1915, p. 108.

12. McDougall, William, *Is America Safe for Democracy?* (New York: Charles Scribner's Sons, 1921); ———, *Ethics and Some Modern World Problems* (New York: G. P. Putnam's Sons, 1924); ———, *The Indestructible Union* (Boston: Little, Brown and Co., 1925).

13. Quoted in *Teaching School Bulletin,* February 1914, p. 160.

14. Fisher, Irving, "What I Think About Eugenics," in Eugenics Society of the USA, *A Brief Bibliography of Eugenics,* 1915, p. 5.

15. Allen, Gar, "A History of Eugenics in the Class Struggle," *Science for the People,* March 1974, p. 33.

16. Hooton, Earnest Albert, *The American Criminal: An Anthropological Study* (Cambridge, MA: Harvard University Press, 1939), pp. 307–9.

17. Godkin, Edwin L., *Problems of Modern Democracy: Political and Economic Essays* (Cambridge: Belknap Press of Harvard University Press, 1966).

18. Croly, Herbert D., *The Promise of American Life* (New York: Macmillan Co., 1909), p. 81.

19. Coolidge, Calvin, "Whose Country Is This?" *Good Housekeeping,* February 1921, pp. 13–14.

20. Bell, A. G., "A Few Thoughts Concerning Eugenics" (an address to the American Breeders Association in Washington, January 1908), *National Geographic,* February 1908, pp. 119–23; Also see A. G. Bell, "How to Improve the Race," *Journal of Heredity,* January 1914, pp. 1–7.

21. Saleeby, Caleb Williams, *The Progress of Eugenics* (New York: Funk & Wagnalls, 1914), p. 89.

22. Sanger, Margaret, "Need for Birth Control in America," in Adolf Meyer, ed., *Birth Control, Facts and Responsibilities* (Baltimore: Williams and Williams Co., 1925), p. 15.

23. Ibid., p. 180; Laughlin, Harry H., *Eugenical Sterilization in the United States* (Chicago: Psychopathic Laboratory of the Municipal Court of Chicago, 1922), p. 15.

24. Hickman, H. B., "Delinquent and Criminal Boys Tested by the Binet Scale," *Teaching School Bulletin,* January 1915, p. 159.

25. Laughlin, Harry H., "Scope of the Committee's Work," Eugenics Record Office, Cold Spring Harbor, New York, 1914.

26. Landman, Jacob H., *Human Sterilization: The History of the Sexual Sterilization Movement* (New York: Macmillan Co., 1932), p. 259.

27. Haller, *Eugenics,* p. 139.

28. Landman, *Human Sterilization,* pp. 80–93.

29. Fisher, Irving, to Charles B. Davenport, "Charles B. Davenport Papers," March 2, 1912.

30. Ross, Edward A., *The Old World in the New: The Significance of Past and Present Immigration to the American People* (New York: The Century Co., 1944), pp. 113, 145, 147, 148, 150.

31. Grant, Madison, *The Conquest of a Continent: or, The Expansion of Races in America* (New York: Charles Scribner's Sons, 1933), p. 53; ———, *Passing of the Great Race: or, The Racial Basis of European History* (New York: Charles Scribner's Sons, 1918), p. 78.

32. Smith, Anthony, *The Human Pedigree* (New York: J. B. Lippincott Company, 1975), p. 233.

33. Laughlin, Harry H., "Analysis of America's Melting Pot," hearings before the house Committee on Immigration and Naturalization, 67th Congress, 3rd session (Washington, D.C.: U.S. Government Printing Office, 1922), p. 755.

34. Davis, James J., "Our Labor Shortage and Immigration," *Industrial Management,* 1923, p. 323.

35. *Congressional Record,* April 8, 1924, p. 5693.

36. Ibid., p. 5872.

37. "Restriction and Immigration," hearings before House Committee on Immigration and Naturalization, 68th Congress, 1st Session (Washington, D.C.: U.S. Government Printing Office, 1924), pp. 767–68.

38. Newman, Horatio H., *Readings in Evolution, Genetics and Eugenics* (Chicago: University of Chicago Press, 1932), p. 531.

39. Beckwith, Jon, "Social and Political Uses of Genetics in the U.S.: Past and Present," *Annals of the New York Academy of Science,* 1976, p. 47.

40. Hitler, Adolf, *Mein Kampf* (Boston: Houghton Mifflin Company, 1943).

41. Dunn, L. C., "Cross Currents in the History of Human Genetics," *American Journal of Human Genetics,* 1962, p. 8.

42. Popenoe, Paul, "The German Sterilization Law," *Journal of Heredity,* July 1934, pp. 257–60.

43. Conversation with L. C. Dunn, October 5, 1970, in Ludmerer, *Genetics and American Society,* p. 133.

44. Ibid., p. 118.

45. Signer, Ethan, "Recombinant DNA: It's Not What We Need," National Academy on Science Forum on Recombinant DNA, March 7–9, 1977.

46. Tomlinson, Eric, "Effect of the New Biologies on Health Care," in Rudolph and McIntire, eds., *Biotechnology,* pp. 68–69; Blakeslee, Sandra, "Treatment for 'Bubble Boy Disease,' " *New York Times,* May 18, 1993, p. C10.

47. Anderson, W. French, "Human Gene Therapy," *Science,* May 8, 1992, p. 812; Henig, Robin Marantz, "Dr. Anderson's Gene Machine," *New York Times Magazine,* March 31, 1991, pp. 30–35, 50.

48. Gorner, Peter, and Ronald Kotulak, "Scientists Criticize Human Gene Therapy," *Chicago Tribune,* September 20, 1990, p. 2; Lehrman, Sally, "Breaking the Code," *San Francisco Examiner,* May 19, 1991, Image p. 23.

49. Kitcher, *The Lives to Come: The Genetic Revolution and Human Possibilities,* pp. 116–17.

50. Ibid., pp. 118–19.

51. Hotz and Maugh, "Biotech: The Revolution Is Already Underway," p. A28.

52. Wheeler, David L., "Few Successes in Gene Therapy," *Chronicle of Higher Education,* July 14, 1995, p. A8.

53. Zimmerman, Burke E., "Human Germ-Line Therapy: The Case for Its Development and Use," *Journal of Medicine and Philosophy,* December 1991, p. 594.

54. Ibid., p. 596.

55. Ibid., pp. 597–98.

56. Lippman, Abby, "Mother Matters: A Fresh Look at Prenatal Genetic Testing," *Issues in Genetic Engineering,* February 1994, pp. 142–43.

57. Ibid.; Cowan, Ruth Schwartz, "Genetic Technology and Reproductive Choice: An Ethics for Autonomy," in Daniel J. Kelves and Leroy Hood, eds., *The Code of Codes: Scientific and Social Issues in the Human Genome Project* (Cambridge, MA: Harvard University Press, 1995), p. 245.

58. Rothstein, Mark A., "Ethical Issues Surrounding the New Technology as Applied to Health Care," in Rudolph and McIntire, eds., *Biotechnology,* p. 204.

59. Kitcher, Philip, *The Lives to Come: The Genetic Revolution and Human Possibilities* (New York: Simon and Schuster, 1996), p. 71.

60. Ibid., p. 69.

61. Thompson, Larry, "Cell Test Before Implant Helps Insure Healthy 'Test-Tube' Baby," *The Washington Post,* April 27, 1992, p. A3.

62. Bodmer, Walter, and Robin McKie, *The Book of Man: The Human Genome Project and the Quest to Discover Our Genetic Heritage* (New York: Scribner, 1994), pp. 234–35.

63. Stamatoyannopoulos, G., "Problems of Screening and Counseling in the Hemoglobinopathies," in Arno G. Motulsky and W. Lenz, eds., *Birth Defects: Proceedings of the Fourth International Conference,* Vienna, Austria, September 2–8, 1973 (Amsterdam: Excerpta Medica, 1974), pp. 268–76.

64. Weiss, Rick, "Discovery of Jewish Cancer Gene Raises Fears of More Than Disease," *The Washington Post,* September 3, 1997, p. A3.

65. Ibid.

66. *Berman v Allan,* 404 A. 2d§80 N.J. 421 (1979), pp. 10–13; Levoy, Gregg, "Wrongful Life: People Now Sue Just for Being Born," *San Francisco Chronicle,* March 25, 1990, This World, p. 12.

67. California Civil Code 43.6 (1991); Indiana Code. Ann. 34-1-1-11 (1990); Minnesota Statutes 145. 424 (1990); 188.130 R.S. Missouri (1989); South Dakota Codified Laws 21-55-(1-4) (1991); Utah Code 78-11-(23-25) (1983). See *Turpin v Sortini,* 643 P. 2d 954 (1982); *Procanik v Cillo,* 478 A. 2d 483 (1983); *Berman v Allan,* p. 14; Levoy, "Wrongful Life: People Now Sue Just for Being Born," p. 12.

68. "Whether to Make Perfect Humans," *New York Times,* July 22, 1982, p. A22.

69. Ibid.

70. Cowan, "Genetic Technology and Reproductive Choice," pp. 249–52; Sandler, Merton, ed., *Amniotic Fluid and Its Clinical Significance* (New York: M. Dekker, 1981), chapter 1; Valenti, Carlo, Edward J. Schutta, and Tehila Kehaty, "Prenatal Diagnosis of Down's Syndrome," *Lancet,* July 27, 1968, p. 220.

71. Cowley, Geoffrey, "Made to Order Babies," *Newsweek,* Winter/Spring 1990, Special Issue, "The 21st Century Family," p. 98.

72. Lehrman, Sally, "The Fountain of Youth?" *Harvard Health Letter,* June 1992, pp. 1–3.

73. Schrof, Joanne M., "Pumped Up," *U.S. News & World Report,* June 1, 1992, p. 55.

74. Werth, Barry, "How Short Is Too Short? Marketing Human Growth Hormone," *New York Times Magazine,* June 16, 1991, p. 47.

75. Memorandum to Gilman Grave, M.D., Chair, ICRS, NICHD, Clinical Research Project Number 91-CH-46, "A Randomized, Double Blind, Placebo-Controlled Clinical Trial of the Effects of Growth Hormone Therapy on the Adult Height of Non Growth Hormone Deficient Children with Short Stature," November 27, 1990; agreement letter from Lilly Research Laboratories to Gordon B. Cutler, Jr., M.D., November 11, 1987.

76. "NIH Hormone Tests with Children Draws Criticism of Group," *Wall Street Journal* (Eastern edition), June 25, 1992, p. A1.

77. Lantos, John, Mark Siegler, and Leona Cuttler, "Ethical Issues in Growth Hormone Therapy," *Journal of the American Medical Association,* February 17, 1989, p. 1022.

78. Werth, "How Short Is Too Short?," p. 15.

79. Parens, Erik, "Autonomous Consumers," *Hastings Center Report,* July/August 1994, p. 3.

80. Zimmerman, "Human Germ-Line Therapy," p. 607.

81. "Changing Your Genes," *The Economist,* April 25, 1992, p. 11.

82. Ibid.

83. Sinsheimer, Robert L., "The Prospect of Designed Genetic Change," *Engineering and Science,* April 1969, pp. 8–13.

84. Quoted in Walter G. Peter III, "Ethical Perspectives in the Use of Genetic Knowledge," *BioScience,* November 15, 1971, p. 1133.

85. Limoges, Camille, "Errare Humanum Est: Do Genetic Errors Have a Future?" in Carl F. Cranor, ed., *Are Genes Us? The Social Consequences of the New Genetics* (New Brunswick, NJ: Rutgers University Press, 1994), p. 114.

86. Ibid., pp. 114–15.

## Five **The Sociology of the Gene**

1. Bouchard, Thomas J., et al., "Sources of Human Psychological Differences: The Minnesota Study of Twins Reared Apart," *Science,* October 12, 1990, p. 223; Dusek, Val, "Bewitching Science," *Science for the People,* November/December 1987, p. 19.

2. Angier, Natalie, "Parental Origin of Chromosome May Determine Social Causes, Scientists Say," *New York Times,* July 12, 1997, p. A18.

3. Brown, David, "Girls May Inherit Intuition Gene from Fathers," *The Washington Post,* July 12, 1997, p. A3.

4. Angier, "Parental Origin of Chromosome May Determine Social Causes, Scientists Say," p. A18.

5. Clarringer, Robert C., Rolf Adolfsson, and Nenad M. Svranic, "Mapping Genes for Human Personality," *Nature Genetics,* January 1996, p. 3.

6. Concar, David, "High Anxiety and Lazy Genes," *New Science,* December 7, 1996, p. 22.

7. Brody, Jane E., "Quirks, Oddities May Be Illnesses," *New York Times,* February 4, 1997, pp. C1–C2.

8. Radford, Tim, "Straight Talk on the Gay Gene: Will Eugenics Come Out of the Closet?" *Guardian* (London), reprinted in *World Press Review,* September 1993, p. 23.

9. Ibid.

10. Holmes, Bob, "A Gene for Boozy Mice," *New Scientist,* September 14, 1996, p. 16.

11. Reiss, David, "Genetic Influence on Family Systems: Implications for Development," *Journal of Marriage and the Family,* August 1995, p. 547.

12. Ibid., p. 551.

13. Ibid., p. 552.

14. Quoted in Smith, *The Human Pedigree,* p. 247.

15. Beckwith, Jon, "A Historical View of Social Responsibility in Genetics," *BioScience,* May 1993, p. 330.

16. U.S. Congress, Office of Technology Assessment, *Mapping Our Genes* (Washington, D.C.: U.S. Government Printing Office, 1988), p. 85; Koshland, Daniel, "Sequences and Consequences of the Human Genome," *Science,* October 13, 1989, p. 189.

17. Kevles, Daniel J., *In the Name of Eugenics: Genetics and the Uses of Human Heredity* (Cambridge, MA: Harvard University Press, 1995), p. 269.

18. Ibid., p. 271.

19. Ibid., pp. 275–83.

20. Koshland, Daniel, "Sequences and Consequences of the Human Genome," *Science,* October 13, 1989, p. 189. See Office of Technology Assessment, *Mapping Our Genes,* p. 84.

21. Koshland, "Sequences and Consequences of the Human Genome," p. 189; ———, "The Human Genome Project: Biological Nature and Social Opportunities," paper presented at the Stanford Centennial Symposium, January 11, 1991.

22. Hubbard, Ruth, and Elijah Wald, *Exploding the Gene Myth: How Genetic Information Is Produced and Manipulated by Scientists, Physicians, Employers, Insurance Companies, Educators, and Law Enforcers* (Boston: Beacon Press, 1993), p. 9.

23. Rich, Alexander, and Sung Hou Kim, "The Three Dimensional Structure of Transfer RNA," *Scientific American,* January 3, 1978, p. 52.

24. Newman, Stuart A., "Genetic Engineering as Metaphysics and Menace," *Science and Nature,* vols. 9/10, 1989, p. 118.

25. Ibid., p. 116.

26. Ibid., p. 118.

27. Beardsley, Tim, "Smart Genes," *Scientific American,* August 1991, pp. 86–95.

28. Beckwith, "A Historical View of Social Responsibility in Genetics," p. 332.

29. Charles, Dan, "Genetics Meeting Halted Amid Racism Charges," *New Scientist,* September 26, 1992, p. 4; "Criminal Error," *New Scientist,* September 26, 1992, p. 3.

30. Wilson, William Julius, *The Truly Disadvantaged: The Inner City, the Underclass, and Public Policy* (Chicago: University of Chicago Press, 1987), p. 22; Magnet, Myron, *The Dream and the Nightmare: The Sixties' Legacy to the Underclass* (New York: William Morrow and Co., 1993), pp. 50–51.

31. Butterfield, Fox, "Disputes Threatens U.S. Plan on Violence," *New York Times,* October 23, 1992, p. A12.

32. Ibid.

33. Toufexis, Anastasia, "Seeking the Roots of Violence," *Time,* April 19, 1993, p. 53.

34. Ibid.

35. Ibid.

36. Ibid.

37. Volkow, Nora D., and Lawrence R. Tancredi, "Neural Substrates of Violent Behavior: A Preliminary Study with Positron Emission Tomography," *British Journal of Psychiatry,* 1987, pp. 668–73; White, B., "Biological Causes for Violent Behavior: Research Could Affect Legal Decisions," *Texas Bar Journal,* 1987, p. 446; Jeffrey, C. Ray, in collaboration with R. V. Del Carmen and J. D. White, *Attacks on the Insanity Defense: Biological Psychiatry and New Perspectives on Criminal Behavior* (Springfield, IL: Charles C. Thomas, 1985), p. 82.

38. Geller, Lisa N., Joseph S. Alper, Paul R. Billings, Carol I. Barash, Jonathan Beckwith, and Marvin R. Natowicz, "Individual, Family, and Societal Dimensions of Genetic Discrimination: A Case Study Analysis," *Science and Engineering Ethics,* vol. 2, no. 1, 1996, pp. 71–74.

39. Ibid., pp. 75–76.

40. Ibid.

41. Cowley, Geoffrey, "Flunk the Gene Test and Lose Your Insurance," *Newsweek,* December 23, 1996, p. 49.

42. Bishop and Waldholz, *Genome;* de Wit, G. W., "Gentechnology, Insurance, and the Future," paper delivered at "II Workshop on International Cooperation for the Human Genome Project: Ethics," Valencia, Spain, November 11–14, 1990.

43. Bishop and Waldholz, *Genome,* pp. 299–300.

44. Berlfein, Judy, "Genetic Testing: Health Care Trap," *Los Angeles Times,* April 30, 1991, p. B2; Kevles, Daniel J., and Leroy Hood, "Reflections," in Kevles and Hood, eds., *The Code of Codes,* p. 324.

45. Pear, Robert, "States Pass Laws to Regulate Uses of Genetic Testing," *New York Times,* October 18, 1997, p. 1.

46. Geller et al., "Individual, Family, and Societal Dimensions of Genetic Discrimination," p. 77.

47. Gostin, Larry, "Genetic Discrimination: The Use of Genetically Based Diagnostic and Prognostic Tests by Employers and Insurers," *American Journal of Law and Medicine,* vol. 17, nos. 1 and 2, 1991, p. 137; Goldstein, Joseph F., and Michael S. Brown, "Genetic Aspects of Disease," in K. Isselbacher, R. Adams, E. Braunwald, R. Peterdorf, and J. Wilson, eds., *Harrison's Principles of Internal Medicine,* 9th ed., vol. 1 (New York: McGraw-Hill, 1980), p. 293.

48. Kevles, *In the Name of Eugenics,* p. 278.

49. Geller et al., "Individual, Family, and Societal Dimensions of Genetic Discrimination," p. 77.

50. U.S. Congress, Office of Technology Assessment, "The Role of Genetic Testing in the Prevention of Occupational Disease" (Washington, D.C.: U.S. Government Printing Office, 1983), p. 37.

51. Brownlee and Siberner, "The Assurances of Genes," *U.S. News & World Report,* July 23, 1990, p. 57.

52. Pear, "States Pass Laws to Regulate Uses of Genetic Testing," p. 1.

53. Geller et al., "Individual, Family, and Societal Dimensions of Genetic Discrimination," p. 78.

54. Nelkin, Dorothy, and Laurence R. Tancredi, "Classify and Control: Genetic Information in the Schools," *American Journal of Law and Medicine,* vol. 17, nos. 1 and 2, 1991, p. 52; Harris, Irving D., *Emotional Blocks to Learning: A Study of the Reasons for Failure in School* (New York: Free Press of Glencoe, 1961), p. 36. See Rose, "What Should a Biochemistry of Learning and Memory Be About?" *Neuroscience,* May 1981, p. 811.

55. Nelkin and Tancredi, "Classify and Control," pp. 55–56; Coles, Gerald, *The Learning Mystique: A Critical Look at "Learning Disabilities"* (New York: Pantheon Books, 1988), pp. 23–24, 43–44; American Psychiatric Association, *Diagnostic and Statistical Manual of Mental Disorders,* 3rd ed. (Washington, D.C.: American Psychiatric Association, 1980). See Biederman, Munir, and Knee, "Conduct and Oppositional Disorder in Clinically Referred Children with Attention Deficit Disorders: A Controlled Family Study," *Journal of American Academy of Child and Adolescent Psychiatry,* September 1987, p. 724; Biederman, Munir, Knee, Habelow, Armentano, Autor, Hoge, and Waternaux, "A Family Study of Patients with Attention Deficit Disorder and Normal Controls," *Journal of Psychiatric Research,* vol. 20, no. 4, 1986, pp. 263–74; Klerman, "The Significance of the DSM III in American Psychiatry," in Andrew E. Skodol, Robert L. Spitzer, and Janet B. W. Williams, eds., *International Perspectives on DSM III* (Washington, D.C.: American Psychiatric Press, 1983); Greenberg and Erickson, "Pharmacotherapy of Children and Adolescents," in Cecil R. Reynolds and Terry B. Gutkin, eds., *The Handbook of School Psychology* (New York: John Wiley & Sons, 1982), pp. 1023, 1025.

56. American Psychiatric Association, *DSM III,* pp. 42–43, 93–95.

57. Coles, *The Learning Mystique,* pp. 70–90; Nelkin, Dorothy, and Laurence R. Tancredi, *Dangerous Diagnostics: The Social Power of Biological Information* (New York: Basic Books, 1989), pp. 125–26. See Volkow, Nora D., and Lawrence R. Tancredi, "Biological Correlates of Mental Activity: Studies with PET," *American Journal of Psychiatry,* April 1991, p. 439; National Institute of Mental Health, U.S. Department of Health and Human Services, "Approaching the 21st Century: Opportunities for NIMH Neuroscience Research," Report to Congress on the Decade of the Brain, (Washington, D.C.: U.S. Government Printing Office, 1988), pp. 25–26.

58. Nelkin and Tancredi, *Dangerous Diagnostics,* pp. 117–21. See Greenberg and Erickson, "Pharmacotherapy of Children and Adolescents," pp. 1030–31.

59. Silver, Lee M., *Remaking Eden: Cloning and Beyond in a Brave New World* (New York: Avon Books, 1997), pp. 4–7.

60. Ibid, p. 9.

61. Randall, John H., *The Making of the Modern Mind: A Survey of the Intellectual Background of the Present Age* (Boston: Houghton Mifflin, 1940), pp. 223–24; Bacon, Francis, "Novum Organum," in *The Works of Francis Bacon,* vol. 4 (London: J. Rivington and Sons, 1778), pp. 114, 246, 320, 325; ———, "The Masculine Birth of Time," in Benjamin Farrington, ed., *The Philosophy of Francis Bacon: An Essay on Its Development from 1603–1609* (Liverpool, England: Liverpool University Press, 1964), pp. 62, 92–93; ———, "Description of the Intellectual Globe," in *The Works of Francis Bacon,* vol. 5 (London: J. Rivington and Sons, 1778), p. 506; Leiss, William, *The Domination of Nature* (Boston: Beacon Press, 1972), p. 58; Merchant, Carolyn, *The Death of Nature: Women, Ecology and the Scientific Revolution* (San Francisco: Harper and Row, 1980), p. 172.

62. Condorcet, Marquis de, *Outlines of an Historical View of the Progress of the Human Mind* (London, 1795), pp. 4–5.

63. Wilson, Edward O., *On Human Nature* (Cambridge, MA: Harvard University Press, 1978), p. 208.

## Six Computing DNA

1. The following works were used in the discussion of the printed page and the Industrial Age, pp. 172–176: Lowe, Donald M., *History of Bourgeois Perception* (Chicago: University of Chicago Press, 1982); Crowley, David, and Paul Heyer, eds., *Communication in History: Technology, Culture, Society* (New York: Longman Publishing Group, 1991); Innis, Harold A., *Empire and Communications,* revised edition (Toronto: University of Toronto Press, 1972); McLuhan, Marshall, *Understanding Media: The Extensions of Man* (Cambridge, MA: MIT Press, 1994); Ong, Walter J., *Orality and Literacy: The Technologizing of the Word* (New York: Methuen & Co., 1982); Eisenstein, Elizabeth L., *The Printing Revolution in Early Modern Europe* (Cambridge: Cambridge University Press, 1983).

2. The following works were used in the discussion of the new language of biology, pp. 176–80: Crowley and Heyer, eds., *Communication in History;* Negroponte, Nicholas, *Being Digital* (New York: Alfred A. Knopf, 1995); Hardinson, O. B., Jr., *Disappearing Through the Skylight: Culture and Technology in the Twentieth Century* (New York: Viking Press, 1989); Heim, Michael, *Electric Language: A Philosophical Study of Word Processing* (New Haven: Yale University Press, 1987); ———, *The Metaphysics of Virtual Reality* (Oxford: Oxford University Press, 1993); Turkle, Sherry, *Life on the Screen: Identity in the Age of the Internet* (New York: Simon and Schuster, 1995); Jones, Steven G., *Cybersociety: Computer-Mediated Communication and Community* (Thousand Oaks, CA: Sage Publications, 1995); Benedikt, Michael, ed., *Cyberspace: First Steps* (Cambridge, MA: MIT Press, 1991); Brook, James, and Iain A. Boal, eds., *Resisting the Virtual Life: The Culture and Politics of Information* (San Francisco: City Lights, 1995); Gelernter, David, *Mirror Worlds, Or, The Day Software Puts the Universe in a Shoebox—: How It Will Happen and What It Will Mean* (Oxford: Oxford University Press, 1991).

3. Weizenbaum, Joseph, *Computer Power and Human Reason: From Judgment to Calculation* (San Francisco: W. H. Freeman, 1976), p. 156.

4. Wiener, Norbert, *The Human Use of Human Beings: Cybernetics and Society* (New York: Avon Books, 1954), p. 26–27.

5. Ibid., p. 35.

6. Rosenblueth, A., N. Wiener, and J. Bigelow, "Behavior, Purpose and Teleology," *Philosophy and Science,* 1943, p. 18.

7. Wiener, *The Human Use of Human Beings,* p. 278.

8. Ibid., p. 25.

9. Rappaport, Roy A., *Ecology, Meaning, and Religion* (Richmond, CA: North Atlantic Books, 1979), p. 169.

10. Dechert, Charles R., "The Development of Cybernetics," in Charles R. Dechert, ed., *The Social Impact of Cybernetics* (New York: Simon & Schuster, 1966), pp. 18–19.

11. McLuhan, *Understanding Media: The Extensions of Man,* pp. 302–3.

12. Ibid., p. 225.

13. Ibid., p. 103.

14. Grene, Marjorie, *The Understanding of Nature: Essays in the Philosophy of Biology* (Dordrecht, Holland: D. Reidel, 1974), p. 68.

15. Gregory, R. L., "The Brain as an Engineering Problem," in William H. Thorpe, and Oliver L. Zangwill, eds., *Current Problems in Animal Behaviour* (Cambridge: Cambridge University Press, 1961), p. 307.

16. Thorpe and Zangwill, *Current Problems in Animal Behaviour,* p. 303.

17. Ibid.

18. Thorpe, W. H., "The Frontiers of Biology," in John B. Cobb, and David R. Griffin, eds.,

*Mind in Nature: Essays on the Interface of Science and Philosophy* (Washington, D.C.: University Press of America, 1977), p. 3.

19. Waddington, C. H., "Whitehead and Modern Science," in Cobb and Griffin, eds., *Mind in Nature,* p. 145.
20. Ibid.
21. Thorpe, "The Frontiers of Biology," p. 6.
22. Dawkins, Richard, *The Blind Watchmaker* (New York: Norton, 1986), p. 112.
23. Dyson, Freeman J., *The Origins of Life* (Cambridge: Cambridge University Press, 1985), p. 6.
24. Grassé, Pierre P., *Evolution of Living Organisms: Evidence for a New Theory of Transformation* (New York: Academic Press, 1977), p. 223.
25. Ibid., p. 224.
26. Ibid.
27. Ibid., p. 225.
28. Ibid., p. 226.
29. Alberts, Bruce, et al., *Molecular Biology of the Cell,* 2nd edition (New York: Garland, 1989), p. 902.
30. Peterson, Ivars, "Computing with DNA-based Computers: Off the Drawing Board and Into the Wet Lab," *Science News,* July 13, 1996, p. 26.
31. Fox, Evelyn Keller, *Refiguring Life: Metaphors of Twentieth-Century Biology* (New York: Columbia University Press, 1995), p. 118.
32. Cantor, Charles, "The Challenges to Technology and Informatics," in Kevles and Hood, eds., *Code of Codes,* p. 106.
33. Ibid., p. 110.
34. Kolata, Gina, "Biology's Big Project Turns Into Challenge for Computer Experts," *New York Times,* June 11, 1996, p. C12.
35. Frankel, Karen A., "The Human Genome Project and Informatics: A Monumental Scientific Adventure," *Communications of the Association for Computing Machinery,* November 1991, p. 40.
36. Cook-Deegan, Robert, *The Gene Wars: Science, Politics and the Human Genome* (New York: W. W. Norton, 1994) pp. 293–94.
37. King, Ralph T., Jr., "Gene Machines: An Eclectic Scientist Gives Biotechnology a Fast Assembly Line," *Wall Street Journal,* May 30, 1995, pp. A1, A5.
38. "DNA Sequencing Software: Balancing Sensitivity, Speed, Flexibility, and Ease of Use," Advertising Supplement, *Science,* June 7, 1996, pp. 1509–10.
39. Ibid., p. 1510.
40. Ibid., p. 1518.
41. King, "Gene Machines," p. A5.
42. Aldhous, Peter, "Managing the Genome Data Deluge," *Science,* October 22, 1993, p. 503.
43. Uehling, Mark D., "Birth of a Molecule," *Popular Science,* February 1992, pp. 74–76.
44. Stepp, David, "Gene Chip Breakthrough," *Future,* March 31, 1997, pp. 58–59.
45. Ibid., p. 59.
46. Carey et al., "The Biotech Century," p. 90.
47. Levy, Steven, "Computers Go Bio: DNA Beats a Pentium Any Day," *Newsweek,* May 1, 1995, p. 63.
48. Bass, Thomas A., "Gene Genie," *Wired,* August 1995, p. 164.
49. Ibid., p. 166.
50. Ibid., p. 168.

## Seven **Reinventing Nature**

1. Hallpike, Christopher R., *The Foundations of Primitive Thought* (Oxford: Clarendon Press, 1979), p. 480.

2. Glacken, Clarence J., *Traces on the Rhodian Shore: Nature and Culture in Western Thought from Ancient Times to the End of the Eighteenth Century* (Berkeley and Los Angeles: University of California Press, 1967), p. 230.

3. Ibid.

4. Quoted in Arthur O. Lovejoy, *The Great Chain of Being: A Study of the History of an Idea* (Cambridge, MA: Harvard University Press, 1936), p. 86.

5. Hoyt, Robert S., *Europe in the Middle Ages*, 2nd ed. (New York: Harcourt, Brace & World, 1966), p. 300.

6. Rank, Otto, *Beyond Psychology* (New York: Dover, 1941), pp. 32–33.

7. Greene, John C., *Science, Ideology, and World View: Essays in the History of Evolutionary Ideas* (Berkeley: University of California Press, 1981), p. 124.

8. Sandow, Alexander, "Social Factors in the Origin of Darwinism," *Quarterly Review of Biology*, September 1938, p. 325.

9. Greene, *Science, Ideology, and World View*, p. 7.

10. West, Geoffrey, *Charles Darwin: A Portrait* (New Haven: Yale University Press, 1938), p. 334.

11. Darwin, Charles, *Charles Darwin's Natural Selection: Being the Second Part of His Big Species Book Written from 1856–1858*, R. C. Stauffer, ed. (Cambridge: Cambridge University Press, 1975), p. 233.

12. Ibid., p. 228.

13. Young, Robert M., "Man's Place in Nature," in Mikulas Teich and Robert Young, eds., *Changing Perspectives in the History of Science: Essays in Honor of Joseph Needham* (Boston: R. Reidel Publishing, 1975), p. 375.

14. Young, Robert M., "Darwinism and the Division of Labour," *Listener*, August 17, 1972.

15. Ibid.

16. Darwin and Wallace, *Evolution by Natural Selection*, p. 119.

17. Bowler, Peter J., "Malthus, Darwin and the Concept of Struggle," *Journal of the History of Ideas*, October/December 1976, p. 645.

18. Quoted in Gertrude Himmelfarb, *Darwin and the Darwinian Revolution* (New York: W. W. Norton, 1959), p. 337.

19. Spengler, Oswald, *The Decline of the West* (New York: Knopf, 1939), p. 373.

20. Himmelfarb, *Darwin and the Darwinian Revolution*, p. 416.

21. Kalmus, Hans, *Regulation and Control in Living Systems* (New York: John Wiley, 1967), p. 151.

22. Ford, John J., "Soviet Cybernetics and International Development," in Charles R. Dechert, ed., *The Social Impact of Cybernetics* (New York: Simon and Schuster, 1966), p. 171.

23. Ibid.

24. Goodwin, Brian, *How the Leopard Changed Its Spots: The Evolution of Complexity* (New York: Simon and Schuster, 1996), p. ix.

25. Ibid., p. xii.

26. Lewin, Roger, *Complexity: Life at the Edge of Chaos* (New York: Macmillan, 1992), p. 16.

27. Goodwin, p. 183.

28. Lewin, p. 137.

29. Ibid., p. 138.

30. Ibid.

31. Ibid., p. 139.

32. Sayre, Kenneth M., *Cybernetics and the Philosophy of Mind* (Highlands, NJ: Humanities Press, 1976), p. 231.

33. Wiener, Norbert, *The Human Use of Human Beings*, p. 38.

34. Trimble, Bjo, *The Star Trek Concordance* (New York: Ballantine Books, 1976), p. 241.

35. Ibid.

36. Ibid.

37. Wiener, Norbert, *The Human Use of Human Beings*, pp. 130–31.

38. Ibid., p. 131.

39. Ibid., p. 140.

40. Masuda, Yoneji, *Managing in the Information Society: Releasing Synergy Japanese Style* (Washington, D.C.: World Future Society, 1980), p. 150.

41. Fjermedal, Grant, *The Tomorrow Makers: A Brave New World of Living-Brain Machines* (New York: Macmillan, 1986), p. 8.

42. Lederberg, Joshua, "Experimental Genetics and Human Evolution," *Bulletin of the Atomic Scientists*, October 1996, p. 6.

43. Huxley, Julian, *Evolution in Action* (New York: New American Library, 1953), p. 31.

44. Ibid.

45. Silver, Lee M., *Remaking Eden: Cloning and Beyond in a Brave New World* (New York: Avon Books, 1997), pp. 249–50.

46. Lederberg, "Experimental Genetics and Human Evolution," p. 9.

47. Dyson, Freeman, "Time Without End: Physics and Biology in an Open Universe," *Reviews of Modern Physics*, vol. 51, p. 449, quoted in Mary Midgley, *Science as Salvation* (London: Routledge, 1992), p. 150.

48. Nelkin, Dorothy, and M. Susan Lindee, *The DNA Mystique: The Gene as a Cultural Icon* (New York: W. H. Freeman and Co., 1995), p. 2.

49. Ibid.

50. Ibid., p. 16.

51. Mumford, Lewis, *Art and Technics* (New York: Columbia University Press, 1952), p. 24.

# Bibliography


Alberts, Bruce, ed. *Molecular Biology of the Cell*, 2nd ed. (New York: Garland, 1989).

American Psychiatric Association. *Diagnostic and Statistical Manual of Mental Disorders*, 3rd ed. (Washington, D.C.: American Psychiatric Association, 1980).

Bacon, Francis. *The Works of Francis Bacon* (London: J. Rivington and Sons, 1778).

Barnaby, Frank. *The Gaia Peace Atlas: Survival into the Third Millennium* (New York: Doubleday, 1988).

Barr, M. W. *Mental Defectives* (Philadelphia: P. Blakiston's Sons and Co., 1904).

Belcher, Brian, and Geoffrey Hawtin. *A Patent on Life: Ownership of Plant and Animal Research* (Canada: IDRC, 1991).

Benedikt, Michael, ed. *Cyberspace: First Steps* (Cambridge, MA: MIT Press, 1991).

Berman, Morris. *The Reenchantment of the World* (Ithaca, NY: Cornell University Press, 1981).

Bishop, Jerry E., and Michael Waldholz. *Genome: The Story of the Most Astonishing Scientific Adventure of Our Time—The Attempt to Map All the Genes in the Human Body* (New York: Simon & Schuster, 1990).

Bodmer, Walter, and Robin McKie. *The Book of Man: The Human Genome Project and the Quest to Discover Our Genetic Heritage* (New York: Scribner, 1994).

Boies, Henry M. *Prisoners and Paupers: A Study of the Abnormal Increase of Criminals, and the Public Burden of Pauperism in the US: The Causes and Remedies* (Freeport, NY: Books for Libraries Press, 1972).

Brockway, Lucile H. *Science and Colonial Expansion: The Role of the British Royal Botanic Gardens* (New York: Academic Press, 1979).

Brook, James, and Iain A. Boal, eds. *Resisting the Virtual Life: The Culture and Politics of Information* (San Francisco: City Lights, 1995).

Burckhardt, Titus. *Alchemy: Science of the Cosmos, Science of the Soul*, William Stoddard, trans. (London: Stuart & Watkins, 1967).

Busch, Lawrence, William B. Lacy, Jeffrey Burkhardt, and Laura R. Lacy. *Plants, Power, and Profit: Social, Economic, and Ethical Consequences of the New Biotechnologies* (Cambridge, MA: Basil Blackwell, 1991).

Cobb, John B., and David R. Griffin, eds. *Mind in Nature: Essays On the Interface of Science and Philosophy* (Washington, D.C.: University Press of America, 1977).

Coles, Gerald. *The Learning Mystique: A Critical Look at "Learning Disabilities"* (New York: Pantheon Books, 1988).

Cranor, Carl F., ed. *Are Genes Us? The Social Consequences of the New Genetics* (New Brunswick, NJ: Rutgers University Press, 1994).

Croly, Herbert D. *The Promise of American Life* (New York: Macmillan Co., 1909).

Crosby, Alfred W., Jr. *The Columbian Exchange: Biological and Cultural Consequences of 1492* (Westport: Greenwood Publishing Co., 1972).

Crowley, David, and Paul Heyer, eds. *Communication in History: Technology, Culture, Society* (New York: Longman Publishing Group, 1991).

Darwin, Charles. *Autobiography of Charles Darwin: 1809–1882*, Nora Barlow, ed. (New York: W. W. Norton, 1958).

———. *The Life and Letters of Charles Darwin,* Francis Darwin, ed. (New York: Appleton, 1887).

———. *Charles Darwin's Natural Selection: Being the Second Part of His Big Species Book Written from 1856–1858,* R. C. Stauffer, ed. (Cambridge: Cambridge University Press, 1975).

———. *The Origin of Species* (New York: Watts, 1929; London: J. M. Dent, 1971).

Darwin, Charles, and Alfred R. Wallace. *Evolution by Natural Selection* (Cambridge: Cambridge University Press, 1958).

Dawkins, Richard. *The Blind Watchmaker* (New York: Norton, 1986).

Dechert, Charles R., ed. *The Social Impact of Cybernetics* (New York: Simon & Schuster, 1966).

de Chobam, Thomas. *Summa Confessorum,* F. Broomfield, ed. (Paris: Louvain, 1968).

Divine, Robert. *American Immigration Policy, 1924–1952* (New Haven: Yale University Press, 1957).

Dyson, Freeman J. *The Origins of Life* (Cambridge: Cambridge University Press, 1985).

Dyson, George B. *Darwin Among the Machines: The Evolution of Global Intelligence* (Reading, MA: Addison-Wesley, 1997).

Eisenstein, Elizabeth L. *The Printing Revolution in Early Modern Europe* (Cambridge: Cambridge University Press, 1983).

Farrington, Benjamin. *The Philosophy of Francis Bacon: An Essay On Its Development from 1603–1609* (Liverpool, England: Liverpool University Press, 1964).

Finberg, H. P. R., and Joan Thirsk, eds. *The Agrarian History of England and Wales, 1500–1640,* vol. 4 (Cambridge: Cambridge University Press, 1967).

Fjermedal, Grant. *The Tomorrow Makers: A Brave New World of Living-Brain Machines* (New York: Macmillan, 1986).

Fletcher, Joseph. *The Ethics of Genetic Control: Ending Reproduction Roulette* (Garden City, NY: Anchor Books, 1974).

Fowler, Cary, and Patrick R. Mooney. *Shattering: Food, Politics, and the Loss of Genetic Diversity* (Tucson, AZ: University of Arizona Press, 1990).

Fox, Evelyn Keller. *Refiguring Life: Metaphors of Twentieth-Century Biology* (New York: Columbia University Press, 1995).

Fox, Michael W. *Superpigs and Wondercorn: The Brave New World of Biotechnology and Where It All May Lead* (New York: Lyons & Burford, 1992).

Frankel, Otto H., and John G. Hawkes, eds. *Crop Genetic Resources for Today and Tomorrow* (New York: Cambridge University Press, 1975).

Gelernter, David H. *Mirror Worlds, Or, The Day Software Puts the Universe in a Shoebox—: How It Will Happen and What It Will Mean* (Oxford: Oxford University Press, 1991).

Glacken, Clarence J. *Traces on the Rhodian Shore: Nature and Culture in Western Thought from Ancient Times to the End of the Eighteenth Century* (Berkeley: University of California Press, 1967).

Godkin, Edwin L. *Problems of Modern Democracy: Political and Economic Essays* (Cambridge: Belknap Press of Harvard University Press, 1966).

Goodwin, Brian. *How the Leopard Changed Its Spots: The Evolution of Complexity* (New York: Simon & Schuster, 1996).

Grant, Madison. *The Conquest of a Continent: or, The Expansion of Races in America* (New York: Charles Scribner's Sons, 1933).

———. *Passing of a Great Race: or, The Racial Basis of European History* (New York: Charles Scribner's Sons, 1918).

Grassé, Pierre P. *Evolution of Living Organisms: Evidence for a New Theory of Transformation* (New York: Academic Press, 1977).

Greene, John C. *Science, Ideology, and World View: Essays in the History of Evolutionary Ideas* (Berkeley: University of California Press, 1981).

Grene, Marjorie. *The Understanding of Nature: Essays in the Philosophy of Biology* (Dordrecht, Holland: D. Reidel, 1974).

Gruber, Howard E., and Paul H. Barrett. *Darwin on Man: A Psychological Study of Scientific Creativity* (New York: E. P. Dutton, 1974).
Guyer, Michael F. *Being Well Born: An Introduction to Heredity and Eugenics* (Indianapolis: Bobbs-Merrill Co., 1916).
Haller, Mark H. *Eugenics: Hereditarian Attitudes in American Thought* (New Brunswick, NJ: Rutgers University Press, 1963).
Hallowell, John H. *Main Currents in Modern Political Thought* (New York: Holt, Rinehart and Winston, 1950).
Hallpike, Christopher R. *The Foundations of Primitive Thought* (Oxford: Clarendon Press, 1979).
Hammond, John L., and Barbara B. Hammond. *The Village Labourer, 1790–1832: A Study in the Government of England Before the Reform Bill* (London: Longmans Green, 1911).
Hardison, O. B., Jr. *Disappearing Through the Skylight: Culture and Technology in the Twentieth Century* (New York: Viking Press, 1989).
Harris, Irving D. *Emotional Blocks to Learning: A Study of the Reasons for Failure in School* (New York: Free Press of Glencoe, 1961).
Hasian, Marouf Arif, Jr. *The Rhetoric of Eugenics in Anglo-American Thought* (Athens: The University of Georgia Press, 1996).
Heim, Michael. *Electric Language: A Philosophical Study of Word Processing* (New Haven: Yale University Press, 1987).
———. *The Metaphysics of Virtual Reality* (Oxford: Oxford University Press, 1993).
Himmelfarb, Gertrude. *Darwin and the Darwinian Revolution* (Garden City, NY: Doubleday, 1959).
Hitler, Adolf. *Mein Kampf* (Boston: Houghton Mifflin Company, 1943).
Hooton, Earnest Albert. *The American Criminal: An Anthropological Study* (Cambridge, MA: Harvard University Press, 1939).
Horsfall, James G., and Albert E. Dimond, eds. *Plant Pathology: An Advanced Treatise,* volume three (New York: Academic Press, 1959).
Howard, Ted, and Jeremy Rifkin. *Who Should Play God? The Artificial Creation of Life and What It Means to the Future of the Human Race* (New York: Dell Publishing, 1977).
Hoyt, Robert S. *Europe in the Middle Ages,* second edition (New York: Harcourt, Brace & World, 1966).
Hubbard, Ruth, and Elijah Wald. *Exploding the Gene Myth: How Genetic Information Is Produced and Manipulated by Scientists, Physicians, Employers, Insurance Companies, Educators, and Law Enforcers* (Boston: Beacon Press, 1993).
Huxley, Julian. *Evolution in Action* (New York: Harper, 1953).
Innis, Harold A. *Empire and Communications,* rev. ed. (Toronto: University of Toronto Press, 1972).
Isselbacher, K., R. Adams, E. Braunwald, R. Peterdorf, and J. Wilson, eds. *Harrison's Principles of Internal Medicine,* 9th ed., vol. 1 (New York: McGraw-Hill, 1980).
Jeffery, C. Ray, in collaboration with Rolando V. Del Carmen and James D. White. *Attacks on the Insanity Defense: Biological Psychiatry and New Perspectives on Criminal Behavior* (Springfield, IL: Charles C. Thomas, 1985).
Jones, Steven G. *Cybersociety: Computer-Mediated Communication and Community* (Thousand Oaks, CA: Sage Publications, 1995).
Jordan, David Starr. *The Blood of the Nation: A Study of the Decay of Races Through Survival of the Unfit* (Boston: American Unitarian Association, 1902).
———. *The Human Harvest: A Study of the Decay of Races Through Survival of the Unfit* (Boston: American Unitarian Association, 1907).
Kalmus, Hans. *Regulation and Control in Living Systems* (New York: John Wiley, 1967).
Kenney, Martin. *Biotechnology: The University-Industrial Complex* (New Haven: Yale University Press, 1986).
Kevles, Daniel J. *In the Name of Eugenics: Genetics and the Uses of Human Heredity* (Cambridge, MA: Harvard University Press, 1995).

Kevles, Daniel J., and Leroy Hood, eds. *The Code of Codes: Scientific and Social Issues in the Human Genome Project* (Cambridge, MA: Harvard University Press, 1992).
Kitcher, Philip. *The Lives to Come: The Genetic Revolution and Human Possibilities* (New York: Simon & Schuster, 1996).
Kloppenburg, Jack R., Jr. *First the Seed: The Political Economy of Plant Biotechnology, 1492–2000* (Cambridge: Cambridge University Press, 1988).
Krimsky, Sheldon. *Biotechnics & Society: The Rise of Industrial Genetics* (New York: Praeger, 1991).
Landman, J. H. *Human Sterilization: The History of the Sexual Sterilization Movement* (New York: Macmillan Co., 1932).
Laughlin, Harry H. *Eugenical Sterilization in the United States* (Chicago: Psychopathic Laboratory of the Municipal Court of Chicago, 1922).
Leiss, William. *The Domination of Nature* (Boston: Beacon Press, 1972).
Lewin, Roger. *Complexity: Life at the Edge of Chaos* (New York: Macmillan, 1992).
Lifton, Robert Jay. *The Protean Self: Human Resilience in an Age of Fragmentation* (New York: Basic Books, 1993).
Lombroso-Ferrero, Gina. *Criminal Man, According to the Classification of Cesare Lombroso* (New York: G. P. Putnam's Sons, 1911).
Lovejoy, Arthur O. *The Great Chain of Being: A Study of the History of an Idea* (Cambridge, MA: Harvard University Press, 1936).
Lowe, Donald M. *History of Bourgeois Perception* (Chicago: University of Chicago Press, 1982).
Ludmerer, Kenneth M. *Genetics and American Society: A Historical Appraisal* (Baltimore: Johns Hopkins University Press, 1972).
Magnet, Myron. *The Dream and the Nightmare: The Sixties' Legacy to the Underclass* (New York: William Morrow and Co., 1993).
Masuda, Yoneji. *Managing in the Information Society: Releasing Synergy Japanese Style* (Washington, D.C.: World Future Society, 1980).
McAuliffe, Sharon, and Kathleen McAuliffe. *Life for Sale* (New York: Coward, McCann and Geoghegan, 1981).
McDougall, William. *Ethics and Some Modern World Problems* (New York: G. P. Putnam's Sons, 1924).
———. *The Indestructible Union* (Boston: Little, Brown and Co., 1925).
———. *Is America Safe for Democracy?* (New York: Charles Scribner's Sons, 1921).
McKim, W. D. *Heredity and Human Progress* (New York: G. P. Putnam's Sons, 1900).
McLuhan, Marshall. *Understanding Media: The Extensions of Man* (Cambridge, MA: MIT Press, 1994).
Merchant, Carolyn. *The Death of Nature: Women, Ecology, and the Scientific Revolution* (San Francisco: Harper and Row, 1980).
Meyer, Adolph, ed. *Birth Control, Facts and Responsibilities* (Baltimore: Williams and Williams Co., 1925).
Miller, R. H., V. G. Pursel, and H. D. Norman, eds. *Biotechnology's Role in the Genetic Improvement of Farm Animals* (Savoy, IL: American Society of Animal Science, 1996).
Motulsky, Arno G., and W. Lenz, eds. *Birth Defects: Proceedings of the Fourth International Conference* (Amsterdam: Excerpta Medica, 1974).
Mulkay, Michael J. *Science and the Sociology of Knowledge* (London: George Allen & Unwin, 1979).
Mumford, Lewis. *Art and Technics* (New York: Columbia University Press, 1952).
———. *The Myth of the Machine: Technics and Human Development* (New York: Harcourt, Brace & World, 1967).
Murphy, Michael P., and Luke A. J. O'Neil. *What Is Life? The Next Fifty Years: Speculations on the Future of Biology* (Cambridge: Cambridge University Press, 1995).
Murphy, Timothy F., and Marc A. Lappé. *Justice and the Human Genome Project* (Berkeley: University of California Press, 1994).

Myers, Norman. *The Sinking Ark: A New Look at the Problem of Disappearing Species* (Oxford: Pergamon Press, 1979).
———. *A Wealth of Wild Species: Storehouse for Human Welfare* (Boulder, CO: Westview Press, 1983).
Nabhan, Gary P. *Enduring Seeds: Native American Agriculture and Wild Plant Conservation* (San Francisco: North Point Press, 1989).
Negroponte, Nicholas. *Being Digital* (New York: Alfred A. Knopf, 1995).
Nelkin, Dorothy, and Susan M. Lindee. *The DNA Mystique: The Gene as a Cultural Icon* (New York: W. H. Freeman & Co., 1995).
Nelkin, Dorothy, and Laurence R. Tancredi. *Dangerous Diagnostics: The Social Power of Biological Information* (New York: Basic Books, 1989).
Newman, Horatio H. *Readings in Evolution, Genetics, and Eugenics* (Chicago: University of Chicago Press, 1921, 1932).
Ong, Walter J. *Orality and Literacy: The Technologizing of the Word* (New York: Methuen & Co., 1982).
Polanyi, Karl. *The Great Transformation: The Political and Economic Origins of Our Time* (Boston: Beacon Press, 1957).
Prescott-Allen, Christine, and Robert Prescott-Allen. *The First Resource: Wild Species in the North American Economy* (New Haven: Yale University Press, 1986).
Raeburn, Paul. *The Last Harvest: The Genetic Gamble That Threatens to Destroy American Agriculture* (Lincoln, NE: University of Nebraska Press, 1995).
Randall, John H. *The Making of the Modern Mind: A Survey of the Intellectual Background of the Present Age* (Boston: Houghton Mifflin, 1940).
Rank, Otto. *Beyond Psychology* (New York: Dover, 1941).
Rappaport, Roy A. *Ecology, Meaning, and Religion* (Richmond, CA: North Atlantic Books, 1979).
Reed, Howard S. *A Short History of the Plant Sciences* (Waltham, MA: Chronica Botanica Co., 1942).
Reynolds, Cecil R., and Terry B. Gutkin, eds. *The Handbook of School Psychology* (New York: John Wiley & Sons, 1982).
Rifkin, Jeremy. *Declaration of a Heretic* (Boston: Routledge & Kegan Paul, 1985).
Rissler, Jane, and Margaret G. Mellon. *The Ecological Risks of Engineered Crops* (Cambridge, MA: MIT Press, 1996).
Rollin, Bernard E. *The Frankenstein Syndrome: Ethical and Social Issues in the Genetic Engineering of Animals* (New York: Cambridge University Press, 1995).
Roosevelt, Theodore. *The Works of Theodore Roosevelt* (New York: Charles Scribner's Sons, 1926).
Rosenfeld, Albert. *The Second Genesis: The Coming Control of Life* (New York: Vintage, 1969).
Ross, Edward A. *The Old World in the New: The Significance of Past and Present Immigration to the American People* (New York: The Century Co., 1914).
Rubenstein, Richard. *The Age of Triage: Fear and Hope in an Overcrowded World* (Boston: Beacon Press, 1983).
Rudolph, Frederick B., and Larry V. McIntire, eds. *Biotechnology: Science, Engineering, and Ethical Challenges for the Twenty-first Century* (Washington, D.C.: Joseph Henry Press, 1996).
Saleeby, Caleb Williams. *The Progress of Eugenics* (New York: Funk & Wagnalls, 1914).
Sandler, Merton, ed. *Amniotic Fluid and Its Clinical Significance* (New York: M. Dekker, 1981).
Sayre, Kenneth M. *Cybernetics and the Philosophy of Mind* (Highlands, NJ: Humanities Press, 1976).
Shiva, Vandana. *Biopiracy: The Plunder of Nature and Knowledge* (Boston: South End Press, 1997).
Silver, Lee M. *Remaking Eden: Cloning and Beyond in a Brave New World* (New York: Avon Books, 1997).
Skodol, Andrew E., Robert L. Spitzer, and Janet B. W. Williams, eds. *International Perspectives on DSM III* (Washington, D.C.: American Psychiatric Press, 1987).
Slater, Gilbert. *The English Peasantry and the Enclosure of Common Fields* (New York: A. M. Kelley, 1968).

Smith, Adam. *An Inquiry Into the Nature and Causes of the Wealth of Nations* (Oxford: Clarendon Press, 1976).
Smith, Anthony. *The Human Pedigree* (New York: J. B. Lippincott Company, 1975).
Smolin, Lee. *The Life of the Cosmos* (Cambridge: Cambridge University Press, 1997).
Spencer, Herbert. *Social Statistics: The Conditions Essential to Human Happiness Specified, and the First of Them Developed* (London: J. Chapman, 1951).
Spengler, Oswald. *The Decline of the West* (New York: Alfred A. Knopf, 1939).
Tate, William E. *The English Village Community and the Enclosure Movement* (New York: Walker, 1967).
Tawney, R. H. *The Agrarian Problem in the Sixteenth Century* (London: Longmans Green, 1912).
Teich, Mikulas, and Robert M. Young, eds. *Changing Perspectives in the History of Science: Essays in Honor of Joseph Needham* (Boston: R. Reidel Publishing, 1975).
Teitel, Martin. *Rain Forest in Your Kitchen: The Hidden Connection Between Extinction and Your Supermarket* (Washington, D.C.: Island Press, 1992).
Thorpe, William H., and Oliver L. Zangwill, eds. *Current Problems in Animal Behaviour* (Cambridge: Cambridge University Press, 1961).
Tomiuk, J. K. Wöhrmann, and A. Sentker, eds. *Transgenic Organisms: Biological and Social Implications* (Basel, Switzerland: Birkhäuser Verlag, 1996).
Trimble, Bjo. *The Star Trek Concordance* (New York: Ballantine Books, 1976).
Turkle, Sherry. *Life on the Screen: Identity in the Age of the Internet* (New York: Simon & Schuster, 1995).
Warshofsky, Fred. *The Patent Wars: The Battle to Own the World's Technology* (New York: John Wiley & Sons, 1994).
Weizenbaum, Joseph. *Computer Power and Human Reason: From Judgment to Calculation* (San Francisco: W. H. Freeman, 1976).
Wertime, Theodore, and James D. Muhly, eds. *The Coming of the Age of Iron* (New Haven: Yale University Press, 1980).
West, Geoffrey. *Charles Darwin: A Portrait* (New Haven: Yale University Press, 1938).
Wheale, Peter, and Ruth McNally, eds. *Animal Genetic Engineering: Of Pigs, Oncomice, and Men* (London: Pluto Press, 1995).
Whelan, William J., and Sandra Black, eds. *From Genetic Experimentation to Biotechnology: The Critical Transition* (Chichester, England: Wiley, 1982).
Whitehead, Alfred North. *Adventures of Ideas* (New York: Macmillan, 1933).
———. *Concept of Nature* (Cambridge: Cambridge University Press, 1920).
———. *Essays in Science and Philosophy* (New York: Philosophical Library, 1947).
———. *Nature and Life* (Cambridge: Cambridge University Press, 1934).
———. *Process and Reality: An Essay in Cosmology* (Cambridge: Cambridge University Press, 1929).
———. *Science and the Modern World* (New York: Macmillan, 1925).
Wilson, Edward O. *The Diversity of Life* (Cambridge, MA: Harvard University Press, 1992).
———. *On Human Nature* (Cambridge, MA: Harvard University Press, 1978).
Wilson, William Julius. *The Truly Disadvantaged: The Inner City, the Underclass, and Public Policy* (Chicago: University of Chicago Press, 1987).
Wiener, Norbert. *The Human Use of Human Beings: Cybernetics and Society* (New York: Avon Books, 1954).
World Commission on Environment and Development. *Our Common Future [The Brundtland Report]* (Oxford: Oxford University Press, 1987).
Worster, Donald, *Nature's Economy: The Roots of Ecology* (San Francisco: Sierra Club Books, 1977).
Zeven, A. C., and A. M. van Harten, eds. *Broadening the Genetic Base of Crops* (Wageningen, Netherlands: Centre for Agricultural Publishing and Documentation, 1979).

# Index

# About the Author

Mr. Rifkin is the author of fourteen books on the impacts of scientific and technological changes on the economy, the workforce, society, and the environment. His books have been translated into fifteen languages and are used in courses in hundreds of universities. Mr. Rifkin is a consultant to heads of state and government officials around the world and speaks frequently before business, labor, and civic forums. He holds a degree in economics from the Wharton School of Finance and Commerce of the University of Pennsylvania, and a degree in international affairs from the Fletcher School of Law and Diplomacy at Tufts University.

Mr. Rifkin is the founder and president of the Foundation on Economic Trends in Washington, D.C.